New Patterns of World Mineral Development

by Raymond F. Mikesell

W.E. Miner Professor of Economics
University of Oregon

BRITISH-NORTH AMERICAN COMMITTEE

Sponsored by
British-North American Research Association (U.K.)
National Planning Association (U.S.A.)
C.D. Howe Research Institute (Canada)

ISBN 0–89068–049–3
Library of Congress Catalog Card Number 79–90054

Published by the British-North American Committee
Printed and bound in the USA
September 1979

Contents

TABLES

Statement of the British-North American Committee to Accompany the Report

The BNAC has observed with growing concern the increasingly political tone of the debate between developed and developing countries—the so-called North-South dialogue. It believes that the underlying community of interest between the two groups of countries has in the process tended to become obscured, and that this applies particularly to the exploitation of natural resources and the evolution of trading relationships. The Committee has therefore sponsored a series of publications in the hope that these will help to restore some greater objectivity of approach.

The first publication was a report on *Mineral Development in the Eighties* issued in 1976. This has been followed by Professor Alasdair MacBean's work, *A Positive Approach to the International Economic Order*, of which Part I, dealing with "Trade and Structural Adjustment," was published in 1978 and Part II, on "Nontrade Issues," is in preparation.

As a further contribution to this series we invited Raymond F. Mikesell, W.E. Miner Professor of Economics at the University of Oregon and well known for his work on minerals, to prepare a report analyzing the evolving patterns of mineral development and the reasons underlying the changes that have taken place.

His report, which we now have pleasure in presenting, addresses the central problem of how adequate supplies of nonfuel minerals for the world economy can be assured over the remainder of this century. It brings out the complexities of modern mining projects and the reasons for the very large increases not only in costs but in lead times that have to be faced. It demonstrates that if developing countries are to secure for their own economies the benefits deriving from exploitation of their mineral resources they need to attract capital and technical and commercial support on a scale that they have no prospect of providing for themselves. It examines the fundamental question of what arrangements can best be made between, on the one hand, developing countries endowed with valuable mineral resources and, on the other hand, those external organizations which have the ability to undertake the necessary work of exploration, appraisal, development, financing, management, and marketing. In short, it discusses the patterns of mineral development that will best serve the interests of all concerned and will be best adapted to the realities of the present situation.

In commending Professor Mikesell's report as a new contribution to thought and action on this most important subject, the undersigned members would like to emphasize and endorse three of his findings.

> First, in considering the most appropriate form of organization to exploit new mineral resources, he brings out the disadvantages inherent in the process that has come to be known as "depackaging" or "unbundling," by which developing countries have sought to acquire particular services from private firms by separate negotiations with a number of them on specific technical requirements. He concludes that the most efficient way of providing these services is to deal with one of the major international mining firms whose skill and experience cover the whole field and to give it the incentive of an equity participation.

Second, he emphasizes, and illustrates with specific examples in Chapter IV of his report, the need and scope for imaginative innovation in devising terms which can accommodate objectives of the host country and of the mining firm that may on the surface appear incompatible. Building further on what has already been achieved might create a new atmosphere of cooperation between the developing and developed worlds, to their mutual benefit.

Third, he suggests that, following high-risk exploration activities that are best undertaken by experienced companies, some involvement by international public lending organizations such as the World Bank can be valuable in bringing an element of order and stability into what may otherwise prove to be a difficult bilateral relationship. This could be highly important in persuading international mining firms and private sources of finance to commit themselves to large new projects, the fruits of which show themselves only over a long period of time.

Members of the Committee Signing the Statement

Chairmen
SIR RICHARD DOBSON
President, B.A.T. Industries Limited

IAN MacGREGOR
General Partner, Lazard Frères & Co.,
Honorary Chairman,
AMAX Inc.

Vice Chairmen
SIR ALASTAIR DOWN
Chairman, Burmah Oil Company

GEORGE P. SHULTZ
President, Bechtel Corporation

Chairman, Executive Committee
WILLIAM I.M. TURNER, JR.
President and Chief Executive Officer,
Consolidated-Bathurst Inc.

Members
J.A. ARMSTRONG
Chairman and Chief Executive Officer,
Imperial Oil Limited

A.E. BALLOCH
Executive Vice President, Bowater
Incorporated

SIR DONALD BARRON
Group Chairman, Rowntree Mackintosh
Limited

CARL E. BEIGIE
President and Chief Executive Officer,
C.D. Howe Research Institute

ROBERT BELGRAVE
Director, BP Trading Company Limited

†I.H. STUART BLACK
Chairman, General Accident
Fire and Life Assurance
Corporation Ltd.

JOHN F. BOOKOUT
President and Chief Executive Officer,
Shell Oil Company

T.F. BRADSHAW
President, Atlantic Richfield Company

SIR GEORGE BURTON
Chairman, Fisons Ltd.

SIR CHARLES CARTER
Chairman of Research and Management
Committee, Policy Studies Institute

J. EDWIN CARTER
Chairman and Chief Executive Officer,
INCO, Ltd.

SILAS S. CATHCART
Chairman & Chief Executive Officer,
Illinois Tool Works, Inc.

HAROLD van B. CLEVELAND
Vice President, Citibank

JAN COLLINS
Chairman, William Collins & Sons

DONALD M. COX
Director and Senior Vice President,
Exxon Corporation

RALPH J. CRAWFORD, JR.
Vice Chairman of the Board, Wells Fargo
Bank

JAMES W. DAVANT
Chairman of the Board and Chief
Executive Officer, Paine, Webber, Jackson
& Curtis Inc.

DIRK DE BRUYNE
Managing Director, Royal Dutch/Shell
Group of Companies

WILLIAM DODGE
Ottawa, Ontario

†No longer a Committee member.

GEOFFREY DRAIN
General Secretary, National Association of Local Government Officers

JOHN DU CANE
Chairman and Managing Director, Selection Trust Ltd.

GERRY EASTWOOD
General Secretary, Association of Patternmakers and Allied Craftsmen

HARRY E. EKBLOM
Chairman and Chief Executive Officer, European American Bancorp

GLENN FLATEN
First Vice President, Canadian Federation of Agriculture

ROBERT M. FOWLER
Chairman, Executive Committee, C.D. Howe Research Institute

MALCOLM GLENN
Executive Vice President, Reed Holdings Inc.

GEORGE GOYDER
Hon. Secretary, British-North American Research Association

HON. HENRY HANKEY
Director, Lloyds Bank International Ltd.

ROBERT HENDERSON
Chairman, Kleinwort Benson Ltd.

ROBERT P. HENDERSON
President and Chief Executive Officer, Itek Corporation

HENDRIK S. HOUTHAKKER
Professor of Economics, Harvard University

TOM JACKSON
General Secretary, Union of Post Office Workers

DONALD P. JACOBS
Dean, Graduate School of Management, Northwestern University

JOHN V. JAMES
Chairman of the Board, President, and Chief Executive Officer, Dresser Industries, Inc.

GEORGE S. JOHNSTON
President, Scudder, Stevens & Clark

JOSEPH D. KEENAN
President, Union Label and Service Trades Department, AFL-CIO

TOM KILLEFER
Chairman of the Board and Chief Executive Officer, United States Trust Company of New York

H.U.A. LAMBERT
Chairman, Barclays Bank International Ltd.

WILLIAM A. LIFFERS
Vice Chairman, American Cyanamid Company

RAY W. MACDONALD
Honorary Chairman, Burroughs Corporation

CARGILL MacMILLAN, JR.
Senior Vice President, Cargill Inc.

J.P. MANN
Deputy Chairman, United Biscuits Holdings Ltd.

A.B. MARSHALL
Chairman, Bestobell Engineering Products Ltd.

DONALD E. MEADS
Chairman and President, Carver Associates

PATRICK M. MEANEY
Group Managing Director, Thomas Tilling Limited

SIR PETER MENZIES
Welwyn, Herts

JOHN MILLER
Vice Chairman, National Planning Association

DEREK F. MITCHELL
Chairman & Chief Executive Officer, BP Canada Limited

MALCOLM MOOS
Hackensack, Minnesota

KENNETH D. NADEN
President, National Council of Farmer Cooperatives

WILLIAM S. OGDEN
Executive Vice President, The Chase Manhattan Bank, N.A.

BROUGHTON PIPKIN
Chairman, BICC Limited

SIR RICHARD POWELL
Director, Hill Samuel Group Ltd.

ALFRED POWIS
Chairman & President, Noranda Mines Limited

J.G. PRENTICE
Chairman of the Board, Canadian Forest Products, Ltd.

LOUIS PUTZE
Director and Consultant, Rockwell International

BEN ROBERTS
Professor of Industrial Relations, London School of Economics

HAROLD B. ROSE
Group Economic Adviser, Barclays Bank Limited

DAVID SAINSBURY
Director of Finance, J. Sainsbury Ltd.

WILLIAM SALOMON
Managing Partner, Salomon Brothers

A.C.I. SAMUEL
Director General, International Group of the National Association of Pesticide Manufacturers

NATHANIEL SAMUELS
Vice Chairman, Kuhn Loeb Lehman Brothers International, Chairman, Louis Dreyfus Holding Company, Inc.

SIR FRANCIS SANDILANDS
Chairman, Commercial Union Assurance Company, Limited

PETER F. SCOTT
President, Provincial Insurance Company, Ltd.

ROBERT C. SEAMANS, JR.
Massachusetts Institute of Technology

LORD SEEBOHM
Chairman, Finance for Industry

THE EARL OF SELKIRK
President, Royal Central Asian Society

LORD SHERFIELD
Chairman, Raytheon Europe International Company

R. MICHAEL SHIELDS
Managing Director, Associated Newspapers Group Ltd.

GEORGE L. SHINN
Chairman, The First Boston Corporation

WILLIAM E. SIMON
New York, New York

ARTHUR J.R. SMITH
President, National Planning Association

LAUREN K. SOTH
West Des Moines, Iowa

E. NORMAN STAUB
Chairman and Chief Executive Officer, The Northern Trust Company

RALPH I. STRAUS
New York, New York

JAMES A. SUMMER
Excelsior, Minnesota

HAROLD SWEATT
Honorary Chairman of the Board, Honeywell, Inc.

SIR ROBERT TAYLOR
Deputy Chairman, Standard Chartered Bank Ltd.

A.A. THORNBROUGH
Deputy Chairman and Chief Executive Officer, Massey-Ferguson Limited

SIR MARK TURNER
Chairman, Rio Tinto-Zinc Corporation Ltd.

JOHN W. TUTHILL
President, The Salzburg Seminar

W.O. TWAITS
Toronto, Ontario

MARTHA R. WALLACE
Executive Director and Director,
The Henry Luce Foundation, Inc.

RICHARD C. WARREN
Consultant, IBM Corporation

GLENN E. WATTS
President, Communications Workers of America, AFL-CIO

W.L. WEARLY
Chairman, Ingersoll-Rand Company

VISCOUNT WEIR
Chairman and Chief Executive, The Weir Group Limited

SIR ERNEST WOODROOFE
Former Chairman, Unilever Ltd.

Preface

My objective in this report is to review the extensive material that has been written on the economics of nonfuel minerals, including that being produced for the minerals policy review currently under way in the U.S. government with its input from private mining companies, as well as a number of studies emanating from the United Nations and the World Bank. I have combined the material contained in many of these studies with the results of my own research on foreign investment in minerals to deal with the central problem of how best to ensure an adequate supply of nonfuel minerals for the world economy over the next two decades or so. I have approached this in the context of certain political realities that are not going to be altered very much by anything we do or say about them, namely, the continued and perhaps increasing role of governments in nonfuel minerals production, both directly as producers or partners in joint ventures with private enterprise, or as promulgators of laws and regulations that have a heavy impact on exploration, investment and production decisions. Much discussion about government intervention has in the past tended to emphasize the policies of the governments of developing countries, but I am not at all sure whether, when everything is considered, future constraints imposed by these governments will prove to be much more worrisome and severe than those of the developed countries.

The major purpose of this study is not to lament or condemn the trend toward increasing governmental participation in the development of mineral resources. (I have tried to be ideologically neutral in dealing with this trend.) Instead, I have explored some of the ways by which the unique resources of international mining companies can continue to be made available to developing countries. This examination considers both foreign equity investment and other means of obtaining capital, technology, skills, and management. While avoiding the conclusion that only international mining enterprises are capable of ensuring adequate supplies of nonfuel minerals, this study brings out clearly the special contributions that they and foreign equity investment can bring to the development of nonfuel minerals resources in the Third World.

My analysis of the conflicts that arise between foreign private investors and developing countries is based on an examination of a considerable number of past and recent mine development agreements. As shown in the study, some of the recent agreements seem to exemplify new approaches to resolving conflicting objectives. Thus, there may be hope that more imaginative agreements will be one avenue for achieving larger participation of international mining firms in the minerals industries of developing countries. Finally, I express a certain amount of optimism that many of these countries are attempting to change the international image of their investment climates and will live up to the new agreements that they negotiate in good faith.

I have not felt it appropriate to draw up a list of specific policy recommendations, although I hope that my report may provide a basis for such recommendations as the British-North American Committee may wish to make.

I would like to express my appreciation for the comments and suggestions received from members of the BNAC, and I am especially grateful for the editorial contributions of Sperry Lea of the Committee staff and F. Taylor Ostrander of AMAX Inc., assistant to Ian MacGregor.

Raymond F. Mikesell
June 1979

Summary of Conclusions

(Each numbered section refers to one of the five chapters of the study.)

I. World reserves of nonfuel minerals are adequate to satisfy demands in the foreseeable future, but the capital required for needed new mining capacity will be enormous, and attracting this capital will present an especially formidable challenge to the developing countries.

Growth of world population and even modest economic growth will generate substantially greater demand for major nonfuel minerals. At even a minimum 3 percent annual increase, this means that present consumption will virtually double by the end of the century.

This study, which concentrates on copper, aluminum, iron, nickel, and tin, agrees with the view of most mineral economists that there will be no shortage of reserves of nonfuel minerals for meeting such requirements. It also supports their view that, given adequate exploration and investment in productive capacity in those locations with the most favorable resource endowments, there should be no substantial rise in real costs of these minerals as a consequence of depletion of reserves.

A substantial portion of the needed reserves, perhaps 40 to 50 percent depending on the mineral, are situated in the developing world, but if those countries are to reap the benefits of their endowments, they will need to attract capital that they do not have as well as large amounts of associated skills and technology.

Another conclusion of the study is that the global need for new capacity to mine and process minerals will be staggering, as will be the volume of capital required to create the new capacity. Global investment in minerals may have to rise from $2 billion per year in recent years to an average of $12.5 billion per year from now to the end of the century, with a substantial acceleration of this needed new investment in the latter half of that period (all values stated in 1977 dollars). In the developing countries alone, the amount of new capital needed may average $5.5 billion a year. This would mean that *external* finance for mining projects and related infrastructure in the developing countries, which has recently averaged less than $1 billion per year, might have to quadruple to an average of $4 billion per year between now and the end of the century, again with a major acceleration in the latter half of this period.

A major barrier to such expansion of mineral producing capacity in a number of developing countries may be their inability to mobilize the required external capital. In many cases, foreign equity participation may provide the only practical answer. Obviously, foreign investors will not be attracted unless the investment climate improves substantially. Currently, governmental controls and excessive taxation and the threat of contract violations probably constitute greater deterrents to foreign investment than does outright expropriation.

II. Traditional patterns of mineral development have been largely overtaken by a marked shift in the ownership of nonfuel minerals industries from international mining firms to host governments; this has been accompanied, especially in the developing countries, by a sharp decline in new private investment in mining and especially in new exploration.

Over the past two decades, ownership of the nonfuel minerals industries in the developing countries has shifted in considerable degree from international mining firms to host governments. This is especially true in the case of copper, bauxite, iron ore, and cobalt, but, even in those cases where substantial host-government ownership is not involved, government controls have increased. These developments have by no means been confined to Third World countries, as the governments of the developed countries have also greatly extended their controls over their resource industries in recent years, and in some cases have assumed ownership as well. Whether or not it is desirable, increased participation in one form or another by governments in their minerals industries is perhaps inevitable.

Up to now, the shift in ownership in the developing countries has not much affected capacity and output at known deposits and existing facilities. However, there has been a notable decline in new mineral investment and an even sharper reduction in new exploration in these areas. All this has coincided with a new emphasis by major mineral development companies in both investment and exploration in the developed countries—notably the United States, Canada and Australia. These trends, if not reversed, threaten to have the most serious consequences for the longer-run growth of mining in developing countries.

III. A comparison of investment decision making, development financing and operations in mining enterprises controlled by majority-owned foreign investors and by host governments reveals the special contributions which international mining firms and foreign equity capital can bring to developing countries if satisfactory and lasting agreements are to be reached.

It is incorrect to assume that because of the shift in the structure of control, foreign equity investment is no longer essential or even important for the growth of the nonfuel minerals industries in developing countries. Up to now, international mining companies have established the mining industries in most countries; indeed, it was they who found and evaluated nearly all of the large ore bodies which government mining enterprises are now exploiting. Third World countries will have to attract substantial external funds in the form of risk capital, and they will also continue to require large inputs of technology, skills and professional management from developed countries for the exploration, development and marketing of their mineral resources. From the analysis in this study, it appears that the most efficient vehicle for transferring these inputs continues to be equity investment by experienced international mining firms. However, both exploration activities and investment expenditures by international mining firms in Third World countries have declined substantially in recent years.

Two standard arguments in favor of state mining enterprises are, first, that such enterprises can hire the necessary foreign inputs for developing a country's mineral resources without paying economic "rents" to an international mining firm; and second, that national mining firms will make sure that the resources are developed in a manner that will maximize the national interest. It is also argued that the various contributions of foreign investment can be "depackaged" by hiring the technology, management and the capital separately without foreign equity interest. This assumes that the specialized inputs of experienced international mining enterprises in the form of exploration, professional management and capacity to mobilize external private debt capital are readily available for purchase separately on the open market. They are not. Moreover, since returns on equity investments have been no higher for mining firms than for firms in other industries, it is doubtful whether a state mining enterprise that undertakes all of the risks itself and hires the technology and other inputs will be able to develop mineral resources at a lower cost than could a competent international firm. Any cost savings are likely to be offset by a loss in productivity. Finally, some governments have been able to negotiate mine development agreements with foreign mining firms that include provisions which assure the development of their mineral resources in a manner consistent with their national interests, while allowing equity participation in the enterprise.

IV. Despite the recent shifts in ownership and control, examination of some recent cases shows that imaginative and innovative mine development agreements can accommodate the different aspirations and contributions of both international mining firms and host countries.

Given the desires of Third World countries for a major stake in the control of their resource industries, it is important that ways be found whereby foreign equity investment, accompanied by politically independent professional management, can operate in a way that will accommodate the requirements of both foreign investors and host governments. Foreign equity investments usually require that the two parties negotiate mine development contracts. No specific formulas to be embodied in these contracts are recommended, as arrangements must be negotiated that satisfy the policies and objectives peculiar to each situation. In some cases, special conditions are also required by potential suppliers of external debt capital.

Some of the more innovative and imaginative mine development contracts that have been negotiated in recent years (of which five are described in Chapter IV), suggest feasible approaches to accommodating the various objectives of the two parties, although these are often considered incompatible. It seems important that contracts be designed to anticipate the possible sources of conflict between foreign investors and host governments that may occur during the life of the contract and to establish modalities for dealing with unforeseen contingencies.

V. Public international financial agencies, especially the World Bank, can play a valuable role by contributing a portion of the external capital and by their presence in the negotiation and conclusion of new-style mineral development agreements.

There have been numerous proposals made for national and international action to promote mineral exploration and investment in the developing countries. A number of programs of both types have been instituted, such as the U.N. Exploration Revolving Fund for Natural Resources, national investment guarantee programs and national and international public financing of mineral investment in the developing countries. Government investment guarantee programs have been effective in promoting investment in the past, and their availability and coverage in the minerals field would be broadened. Bilateral investment treaties are believed to have had some value, and international codes of conduct involving OECD countries and those developing countries that want to attract foreign investment and are willing to abide by their covenants may serve to encourage foreign investors.

International public institutions such as the World Bank and the regional development banks may be neutral with respect to whether they loan to private or government mining enterprises in developing countries; they are concerned mainly with the technical, managerial and economic soundness and financial viability of the projects in which they participate. These international institutions have made relatively few loans to the minerals sector in developing countries. They could play a larger role in encouraging the flow of both equity and debt capital to mining enterprises in which foreign firms were participating. Although its resources are not large, the International Finance Corporation, which is empowered to take equity positions in mining enterprises in developing countries, is particularly well suited to promote confidence in investing in mining projects. However, mining development in which international financial agencies are likely to participate must be preceded by high-risk exploration activities which experienced international mining companies are best equipped to undertake and for which few effective alternatives have been found.

I. The Adequacy of World Supplies of Nonfuel Minerals to Meet World Requirements and the Cost of Creating New Capacity

This study is concerned with the creation of productive capacity to meet projected demands for nonfuel minerals over the remainder of this century. Since a short report cannot review demand, supply and capacity requirements for all minerals, special attention is given to five basic industrial minerals: copper, aluminum, iron ore, nickel, and tin.

In this chapter, the possible growth of world demand for these major minerals is first assessed, followed by an analysis of their supply availabilities and then of the capital required to bring in the needed new capacity to meet projected demands. Finally, estimates are developed of capital requirements if these countries can develop their mineral potentials and how much must come from external financing.

MINERAL POLICIES AND OBJECTIVES

To begin, a few general comments are appropriate.

It is surely in the interest of all countries that world production of minerals increases broadly in relation to the growth of demand, and that costs be minimized by an efficient allocation of investment expenditures—including exploration costs—among the world's mineral regions. Those sources of supply should not, however, be so concentrated in one country or region as to render the world vulnerable to severe supply disruption or curtailment arising from economic, political or natural causes.

Of course, both consumers and producers of minerals share other interests, such as the avoidance of sharp fluctuations in prices that provide no economic benefits, protection of the environment, and the employment of extractive methods that will not waste the earth's nonrenewable resources.

Individual nations and groups of nations with common political and economic goals, such as the Organization for Economic Cooperation and Development countries, have special objectives with respect to mineral supplies, which they may pursue individually or collectively. Among the more important of these are the security of supplies of important minerals, and protection from the activities of cartels that might raise prices well above competitive levels through restrictions on output or export. Several types of defenses may be employed, including import restrictions to maintain or expand domestic production; the use of economic or strategic stockpiles; or the encouragement of production in relatively "safe" areas. Of the various approaches to security of supply, promoting domestic self-sufficiency by means of import restriction is certainly the least desirable from the standpoint of the interests of both domestic consumers and foreign producers. Nations should rely on imports for their mineral requirements that cannot be supplied domestically at world competitive prices. A case can be made for government stockpiles of those minerals for which the threat of supply disruption from political action is serious, e.g., petroleum from the Middle East, or the platinum-group metals supplied almost entirely by South Africa and the USSR.

For many minerals, however, the objectives of the OECD countries can be met by widely ranging exploration and development of resources throughout the world. Such diversification of sources reduces the risks of both supply disruption arising from political events, e.g., the interruption of rail traffic in Africa, and unwarranted price rises promoted by cartels. Worldwide exploration and the free flow of capital and technology tend to promote this diversification of supply sources; in addition, they help to ensure the growth of economically productive capacity in line with the growth of demand for minerals while achieving the best use of the world's resources of minerals and capital.

HOW MUCH DO WE NEED?

Two major factors determining the demand for minerals are the size of the population and its relative standard of living. The world seems headed for a doubling of its present population by or soon after the year 2000. However, by present trends, two-thirds of the world's *increased* population will presumably live in the 100 or so countries now described as "developing" or "less developed." If we assume any improvement at all in the average material condition of the world's peoples over the next 20 to 30 years, the largest proportion of the improvement is likely to occur in the developing world.

The implication of these two factors for the demand for minerals is simple: it means steady increases in mineral use. In discussing this "unrelenting pressure" for additional supplies of minerals, a World Bank publication stated: "Population growth and improvement in individual income translate into the one basic imperative: *more!*"[1]

---

1 Rex Bosson and Benison Varon, *The Mining Industry and the Developing Countries* (Oxford University Press, 1977), p. 22.

TABLE I-1: GROWTH RATES IN WORLD DEMAND FOR SELECTED NONFUEL MINERALS*
(Annual averages over periods indicated)

	Historical	Projected
	Averaged over recent 15- to 25-year periods	**Range of projections for 15- to 25-year periods to 1990 or 2000**
Aluminum	7.3%	3.0–6.7%
Refined copper	3.9	2.1–4.0
Iron ore	3.6	2.1–3.2
Nickel	6.5	2.1–5.1
Tin	1.0	1.3–1.7

* Covers non-Communist countries.
Source: Appendix Table A–1.

World population is growing about 2 percent per year. Given any increase in average standards of living of the world's peoples, per capita world demand for minerals is likely to grow by at least 3 percent per year. Appendix A discusses a range of demand forecasts for the five major nonfuel minerals under consideration. These projections, summarized in Table I–1, vary considerably, and there is inevitably a large margin of error.

These growth projections range from 1.3 to 3.0 percent on the low side and from 1.7 to 6.7 percent on the high side, depending on the mineral. If we take a median range of, say, 3 to 5 percent annual growth in world demand for the rest of this century, this means that by the middle of the decade of the '80s, world productive capacity for minerals would have to increase—depending upon the mineral in question—by some 25 to 40 percent over existing levels and, by the year 2000, by between 90 and 190 percent.

There are few minerals for which growth in this range would not before long surpass existing capacity and ultimately exceed even our existing knowledge of mineral deposits. Thus, it is obvious that growing world demand will require expensive new exploration and much successful discovery as well as the creation of large new units of physical capacity, all requiring vast amounts of capital, as discussed at the end of this chapter.

This sets the stage from the side of demand.

IS THE SUPPLY OF MINERALS ADEQUATE?

A review of the supply availability of major nonfuel minerals raises no cry of alarm for a critical shortage relative to demand. The studies summarized in Appendix B conclude that, in the absence of a general holocaust, world supplies of nonfuel minerals should be able to grow roughly in relation to the increase in demand.

Although real costs may rise, depending in part upon the geographical location of new productive capacity, most mineral economists believe that depletion of world mineral resources is not likely to be a problem in the sense of substantially raising the cost of finding and processing nonfuel minerals during the remainder of the present century.[2] This conclusion is based on the expectation that the cost-decreasing effects of new technology will keep pace with the cost-increasing effects of having to exploit lower-grade deposits. However, environmental regulations and high energy costs have been increasing real costs of production significantly. Hopefully, the rise in costs resulting from these factors will not continue at the same rate as it has during the 1970s.

Our review of the relevant material shows that even though in 1978 the nonfuel mining industry was generally depressed and faced with both low prices in relation to costs and overcapacity in relation to demand, existing mineral capacity must ultimately expand to meet future requirements. Moreover, because of the very long lead times in modern mineral development, some investment for increased capacity must be initiated soon if supply shortages are not to be encountered even during the 1980s.

Further details on the additional productive capacities for each of five metals needed to meet the projected world demands through the year 2000, and on the amount of investment needed to create these capacities, are given in Appendix C.

The quantitative results of these analyses should be regarded more as an indication of the general magnitude of productive capacities and investment required than as forecasts. Even if these estimates are in error by as much as 25 percent in either direction, they are generally in line with those prepared by economists at the World Bank and the United Nations and by industry experts. Policy

---

2 John E. Tilton, *The Future of Nonfuel Minerals* (Washington, D.C.: The Brookings Institution, 1977), pp. 91–92.

Only in the case of mineral *fuels*, particularly petroleum and uranium, have fears been widely expressed for the serious consequences of depletion over the next quarter century, and even here there is widespread disagreement among students of the subject. The most controversial issue has to do with the relationship between world demand and the existing and potential world reserves of petroleum, a subject which will not be discussed in this study. In the case of uranium, the supply problems of which are most closely related to those of nonfuel minerals, world reserves (estimated at 2.2 million mt in 1976) appear to be adequate for projected consumption to the end of the present century, by which time successful development of the breeder reactor and the reprocessing of fuels should assure adequate supplies for the indefinite future.

determination does not require a high degree of accuracy in such projections, but an idea of the magnitudes involved.

ARE SUPPLIES FROM THE DEVELOPING COUNTRIES IMPORTANT?

It appears that the OECD countries plus South Africa are potentially self-sufficient in the major nonfuel minerals. Why then is there such concern about promoting the productive capacity of nonfuel minerals in the developing countries?

Several factors contribute to an answer.

First, the development of the mineral potential of those developing countries with important reserves will contribute substantially to their own economic welfare. For some countries such as Bolivia, Chile, Jamaica, Papua New Guinea (PNG), Zaire, and Zambia, over half of their export income is now derived from nonfuel minerals. At the same time, all developing countries will benefit greatly as consumers of minerals if adequate supplies help keep down their prices.

In the case of copper, the annual growth rate of consumption in developing countries over the 1975–2000 period is estimated to be about 10 percent as against only 3 percent for developed countries. Higher rates of growth in consumption in the developing countries as compared with those in the developed countries will probably be the case for bauxite, iron ore, lead, nickel, manganese, tin, and zinc, among others.

Second, for some minerals, the reserves within the OECD countries are inadequate or of very low quality. These include such important examples as bauxite, tin, cobalt, chrome, manganese, and the platinum group. Substantial reserves and production of the last three minerals exist in South Africa, a country facing political uncertainties.

Third, there are strategic reasons for diversifying supply sources of minerals where major dependence is on one or two countries. The 1978 disruption of cobalt supplies from Zaire, which produces half the cobalt output of the non-Communist world, demonstrated the threat of excessive reliance on one country.

Diversification of sources of mineral supplies takes on special significance in the light of the heavy dependence of the OECD countries on South African mineral production. About 40 percent of the non-Communist supplies of chromium and manganese and over 80 percent of the platinum-group metals come from South Africa. Loss of these supplies through political difficulties in South Africa would render the OECD countries heavily dependent for these minerals upon the Soviet bloc countries which, in any case, probably could not make up the supply deficits. Rhodesia is another important producer of chromium, and the outlook for uninterrupted production there is even more uncertain than the outlook for South Africa. Even if current output continues to be available from South Africa without interruption, there might not be an adequate level of investment to meet future world demand for these two metals. Substantial reserves of chromium and manganese exist in the market-economy countries outside South Africa and Rhodesia, and more intensive exploration might uncover additional reserves of the platinum-group metals as well. In addition, there are substitutes for all three of these metals in many uses, but adequate stockpiles would be required to provide lead time to introduce them.

Finally, the failure to discover new reserves and expand production of minerals in the developing countries over the next two decades might necessitate the exploitation of lower-grade reserves in the OECD countries of such basic minerals as copper, iron ore, lead, nickel, and zinc, with resultant higher costs.

In the case of copper, it is known that some developing countries have large relatively high-grade copper reserves that are potentially capable of being extracted at costs significantly lower than those for new mines in the United States and Canada. Much the same can be

TABLE I-2: WORLD CAPITAL EXPENDITURES FOR ESTIMATED ADDITIONAL CAPACITY REQUIREMENTS FOR SELECTED MINERALS, 1977-2000[1]
(Billions of 1977 dollars)

	Total[2]		In Developing Countries[2]	
Bauxite	$ 6.9		$ 5.2	
Alumina	24.4		6.1	
Aluminum	76.6		17.6	
Subtotal		107.9		28.9
Copper	58.0[3]		29.0	
Nickel	12.5		5.0	
Iron ore	98.2		31.4	
Tin	1.7		1.4	
Total	$278.3		$95.7	

1 Excludes capital outlays for pollution abatement and exploration.
2 See text of Appendix C for assumptions and calculations.
3 From Appendix Table C-3.
Note: All capital expenditures for years other than 1977 were converted to 1977 dollars by applying the U.S. implicit GNP deflator.

said for nickel and iron ore, abundant high-grade reserves of which exist in the developing countries where potentially they could be produced at a lower cost than in new mines in North America.

However, domestic political and other conditions in the developing countries frequently add sufficiently to total costs to negate their cost advantage based on ore grade alone. High interest rates associated with political uncertainties as well as infrastructure requirements that are substantially greater than in developed countries are two of the principal factors tending to offset for the developing countries the physical advantages of their ore grade.

Some regional cost differences are discussed further in Appendix D.

CAN WE AFFORD THE COSTS OF CREATING NEW SUPPLIES TO MEET RISING WORLD DEMAND?

Estimated capital requirements for meeting projected needed additions to productive capacities for the five nonfuel minerals total, in the aggregate, approximately $278 billion (in 1977 dollars) for the 23 years to the end of the century (see Table I–2) based on estimates given in Appendix C.[3]

These estimates of required world capital expenditures do not include capital outlays for pollution abatement for existing capacity, nor do they include outlays for further exploration. In addition, expenditures will be needed for increased productive capacities for a number of other important nonfuel minerals such as chromium, manganese, molybdenum, tungsten, platinum-group metals, silver, lead, and zinc. (A very rough estimate of the capital requirements for meeting additional lead and zinc capacities alone over the 1977–2000 period is $12 billion.) If capital requirements for other nonfuel minerals were included, plus outlays for pollution abatement and exploration through the year 2000, the total might exceed $300 billion (again in 1977 dollars), or an average of $12.5 billion per year

---

3 The estimates of additional capacity requirements for meeting projected increases in consumption to the year 2000 are based on demands for primary metal. In most cases, it is assumed that the proportion of consumption supplied from secondary sources or recycling will remain constant. If the proportion of total supplies from secondary sources should increase, additional primary capacity requirements would decline accordingly. It has also been assumed that additional consumption requirements for the market economies will be met from non-Communist sources for the commodities indicated. This is unlikely to be the case for each of the commodities discussed. For example, Poland may become an increasingly important copper exporter, but in amounts not likely to alter market-economy capacity requirements significantly.

TABLE I-3: AN ESTIMATE OF AVERAGE ANNUAL FINANCING REQUIREMENTS FOR NONFUEL MINERALS, 1977-2000

	Billions of 1977 Dollars	Comments
In all countries		
(A) Total	$12.5	Derived from the $278 billion estimate for 1977-2000 total in Table I-2, supplemented to cover other minerals and costs for pollution abatement and exploration to total $300 billion for the period.
In developing countries		
(B) Total	5.5	Assumes the developing countries account for 40% of total (= $5 billion) plus infrastructure costs of $0.5 billion.
(C) Of which external financing*	4.1	Assumes 3/4 of total developing country financing must come from external sources.

* By comparison, the World Bank study by K. Takeuchi, G. Thiebach and J. Hilmy, "Investment Requirements in the Nonfuel Minerals Sector of the Developing Countries," *Natural Resources Forum* (April 1977), estimates the average annual external financing for warranted capacity expansion of nonfuel minerals production in the developing countries at $4.6 billion for 1976-80 and $7.3 billion for 1981-85 (all in 1975 dollars).

over the 1977–2000 period. On the basis of their reserve potential, perhaps as much as 40 percent, or $5 billion annually, of these capital outlays might be reasonably considered as the probable share for the developing countries. To this amount must be added a substantial sum for infrastructure in the form of highways, railroads, communities, and so forth, which is also not included in the estimates of capital requirements for productive capacity. This might add at least another half billion dollars to the annual capital requirements for the developing countries, to total $5.5 billion, of which an estimated three-fourths, or $4.1 billion, must be provided externally. (The steps in the development of this estimate are recapitulated in Table I-3.)

Sources of Financing

The total capital required to meet projected world demand for minerals is likely to be available, but since reserves of many metals exist in both developed and developing countries, we cannot be sure of the distribution of future additions to capacity between these two groups of countries. Our estimate of $5.5 billion annually (in 1977 dollars) for "warranted" capital expenditures for nonfuel minerals in the developing countries is a staggering amount, especially in view of the fact that most of this financing would need to come from investors in the industrialized countries. In 1976, the aggregate flow of net financial resources to the nonoil developing countries from the developed countries, from OPEC countries, and from multinational agencies was nearly $60 billion. However, the vast bulk of the public, private and external capital flows to the developing countries is used for financing public utilities, social programs, manufacturing, and the oil deficit, not minerals. Moreover, foreign capital required for metal producing capacity is concentrated in the important mineral exporting countries which represent less than one-fourth of the number of developing countries of the world.

Although there are no data known to the author on the annual volume of aggregate investment in the primary metals industries in developing countries over the past few years, external capital flows to these industries during the 1970s have been only a fraction of the $5.5 billion projected annually for the 1977–2000 period. Over the period 1973–76, net U.S. direct investment capital flows to the mining and smelting industries in the developing countries averaged less than $200 million per year. Over the five-year period ending June 1976, World Bank and International Development Agency (IDA) loans to the nonfuel minerals sector averaged only $50 million a year. In recent years, most of the external capital for the mining industries of the developing countries has taken the form of intermediate commercial bank loans, credits from mining equipment companies supplied or guaranteed in part by government credit institutions, and loans from consumers. Many of these credits have been associated with equity investment by international mining companies. On the basis of a very rough review of the financing of known projects, the author has concluded that external financing of mining projects in the developing countries has averaged less than $1 billion per year over the past five years. Assuming that one-fourth of the projected $5.5 billion per year can be supplied from internal sources, how can the recent $1 billion level of external financing be quadrupled over the next two decades to supply the remaining three-fourths or $4+ billion per year shown in Table I–3?[4]

Countries with growing economies and good export earning prospects may have little difficulty in financing the development of their mineral resources. Brazil, Venezuela and the oil surplus

4 As the footnote to Table I–3 reveals, a World Bank estimate foresees an even larger requirement for external financing of Third World mineral development than is estimated here.

economies of the Middle East obviously fall into this category. There are some others, but everyone would not draw up the same list. For other developing countries, external capital for mineral development will depend heavily upon the willingness of international mining companies to undertake exploration and equity investment without which external financing is unlikely to be forthcoming in sufficient quantities. Some governments will be able to borrow directly in the international financial markets on their own guarantee, but several important mineral economies, including Peru, Zaire and Zambia, are in such dire financial circumstances that large net borrowing by their governments in international markets appears to be quite unlikely for the foreseeable future.

Thus, the real question of adequacy centers on whether the developing countries with significant mineral endowments can mobilize the external capital required to develop them for their own and general benefit. Several aspects of this question are discussed in the chapters that follow.

II. Shifting Patterns of Foreign Ownership and Investment

The ownership structure of several important nonfuel minerals industries in the developing countries has changed substantially over the past 30 years and particularly during the past decade. Specifically, ownership has shifted from foreign (usually private) mining companies to national (usually government-owned) enterprises either as a consequence of expropriation by the host government or occasional voluntary divestiture. This chapter describes these developments as well as recent changes in the level of foreign investment and exploration activity in the developing countries, and assesses their impact on future world supplies.

The shift of ownership from the international mining companies to host governments has provoked two apparently opposing types of concerns. One is that the developing countries will be unable to attract sufficient technology and financial capital, including high-risk equity capital, to develop their mineral reserves fast enough to ensure adequate output of minerals without substantial increases in real costs and prices. The other concern, which is frequently voiced by officials of international mining firms, is that national ownership and control of mineral resources in the developing countries will lead to excess capacity and thus to lower world prices that then fail to cover the full economic costs of production. This, of course, impairs the profitability of private mining firms and reduces incentives to invest, even in the industrialized countries. Moreover, private executives fear that government mining enterprises will not cut back production in response to declines in world demand and prices, but will continue to produce even if their direct costs are not covered. These two concerns need not be wholly inconsistent, since the creation of overcapacity may be only a matter of poor timing of new capacity in relation to the growth of demand.

RECENT CHANGES IN THE OWNERSHIP STRUCTURE IN DEVELOPING COUNTRIES

Before considering the implications of the changed pattern of ownership and control, it may be useful to consider recent developments in ownership in the five basic minerals, as well as several others.

Copper

The most drastic shift has occurred in the copper industry. According to Sir Ronald Prain, a noted analyst of this industry, at the beginning of the 1960s, governments had an interest in only 2.5 percent of the copper producing capacity of the market economies; but, by 1970, about 43 percent of copper producing capacity in these countries was owned in whole or in part by governments.[1] Orris Herfindahl estimated that in 1947 four private mining firms accounted for about 60 percent of world copper output (excluding the USSR) and eight firms accounted for 77 percent. By 1956, these percentages had declined to 47 and 70 percent, respectively.[2] In 1974, the four largest private copper producers—Kennecott, Newmont, Phelps Dodge, and Rio Tinto Zinc—had a majority ownership in less than 19 percent of mine copper output of the market economies, and 10 large privately owned companies had a majority interest in less than 35 percent.[3] Ten other privately owned companies were majority owners of an additional 10 percent of the copper output of the market economies.

Examination of the present situation in the developing countries shows that in 1977 about 60 percent of their primary copper output was produced by majority-owned government enterprises. They controlled the industry in Chile, Zambia and Zaire as well as Peru's second largest copper producer, Cerro de Pasco. In Mexico and the Philippines, foreign copper companies have been forced to reduce their participation to less than majority ownership, and private national ownership dominates the copper industry in both countries. In the Philippines, the shift of foreign ownership from majority to minority equity participation in the mining industry has been accomplished without substantial loss of access to foreign technical and managerial inputs, and the copper industry in that country appears to be making progress. Also, in the Philippines, long-term management contracts may be held by foreign investors with a minority position. On the other hand, the Mexicanization policy requires both majority national ownership and national control so that foreign minority equity holders have not been able to obtain management contracts which ensure their control of operations.

---

1 Sir Ronald L. Prain, *Copper: The Anatomy of an Industry* (London: Mining Journal Books, 1975), pp. 22–23.

2 Orris C. Herfindahl, *Copper Costs and Prices: 1870–1957* (Baltimore: Johns Hopkins University Press for Resources for the Future, 1959), p. 165.

3 The 10 companies are Anaconda, ASARCO, Cyprus, Duval, International Nickel, Kennecott Copper Corporation, Mt. Isa (owned by MIM Holdings of Australia), Newmont, Phelps Dodge, and Rio Tinto Zinc (RTZ). RTZ controls the Bougainville mine in Papua New Guinea (PNG) through its Australian subsidiary, Conzinc Riotinto of Australia, the Lornex mine (Canada) and the Palabora mine (South Africa). ASARCO has a 49 percent interest in Mt. Isa through MIM holdings; with the completion by the Southern Peru Copper Company of the Cuajone mine in Peru, ASARCO—the majority partner in SPCC—became the second largest private copper producer in the world.

To summarize, in the market economies (including Yugoslavia), about 20 private companies with 44 percent of total world copper output still have a majority interest while another 34 percent is produced by majority-owned government enterprises in eight important copper producing countries. Most of the remaining 22 percent is produced by a fairly large number of privately owned companies, some of which have a minority government participation.

The bulk of the mine copper produced by majority-owned government enterprises is smelted by the same firms. In Chile, Peru, Zaire, and Zambia, they refine a substantial portion of their output. Except for the refinery at Ilo, Peru, the refining and smelting facilities were established mainly by international mining companies before their properties were expropriated.

It should be noted that several countries that have expropriated foreign-owned copper mining properties are permitting *new* majority-owned foreign investment in copper mining. Such is the case, for example, in Chile, Peru and Zaire. In some countries, the restriction of foreign investors to a minority position in mining enterprises has undoubtedly deterred exploration and development of minerals. This is particularly true in Mexico where the Mexicanization program has reduced exploration and development in such important minerals as lead, zinc and sulphur, as well as copper.

Iron Ore

Well over half of the iron ore production in the developing countries is controlled by state mining enterprises, but most of their output is consumed domestically. In 1975, the iron ore industries of Venezuela and Peru, both important exporters, were expropriated, including the iron ore properties of U.S. Steel and Bethlehem Steel in Venezuela and those of the U.S.-owned Marcona Company in Peru. In Brazil, the state enterprise, Cia. Vale do Rio Doce (CVRD), dominates the iron ore industry, but private iron ore companies also operate; one of these has formed a joint venture with the Hanna Mining Company for a large expansion of production. The Brazilian iron ore industry, which appears to be making progress, is likely to become an increasingly important exporter. Liberian iron ore production, which is largely exported, continues to be under the control of international mining companies, although there are also joint ventures with the government.

India is an important iron ore exporter and has entered into joint ventures with Japan and more recently with Iran for the development of its iron ore reserves. The Indian mining industry has not, for the most part, been developed by international mining firms as the Indian government participates heavily in both the production and export of minerals. However, a substantial proportion of India's iron ore output is produced by private firms in Goa.

Because the bulk of the iron ore exports are from developed countries (notably Australia and Canada), and the OECD countries are fairly self-sufficient in iron ore production, any future expropriations in the developing countries are likely to have little impact on the availability of iron ore supplies for the OECD countries.

Bauxite

Expropriations in the bauxite industry include Alcan's producing facility in Guyana in 1971 and the previous takeover of the bauxite properties of an Alcan subsidiary in Guinea in 1961.[4] Following the imposition in 1974 of new bauxite levies in violation of the Jamaican government's agreements with the foreign companies, new agreements were reached by Kaiser Aluminum and Reynolds whereby the government acquired 51 percent of their bauxite properties, but management remains in their hands. Several other joint ventures involving national governments and foreign mining enterprises produce bauxite and alumina. Nevertheless, in 1974, about two-thirds of the world's bauxite production was controlled by eight private international firms, six of which—Alcoa, Kaiser, Reynolds, Alcan, Alusuisse, and Pechiney— are also important producers of aluminum metal. Although national governments are likely to continue to seek additional revenues from bauxite production and to press for downstream processing into alumina and aluminum (wherever electric power is adequate), additional expropriations do not appear to be an important threat in the bauxite industry.

Nickel

Virtually no expropriations have occurred in the nickel mining industry since Castro nationalized the Cuban nickel industry in 1960. Over 85 percent of nickel production in the market economies is accounted for by five international mining companies, and well over half the nickel output is produced in four developed countries: Canada, Australia, Greece, and South Africa. Among the developing countries, about half of the nickel output comes from the French Territory New Caledonia, where there is some French government participation. A government-owned enterprise in Indonesia con-

---

4 In 1962–63, Alcan's concession at the large Boke deposit in Guinea was turned over to a joint-venture enterprise 51 percent owned by Halco Mining Inc., a subsidiary of Harvey Aluminum, with 49 percent owned by the government of Guinea. In 1968, Halco divided its 51 percent share among a consortium of Western firms which included Alcan.

trols about 5 percent of the nickel ore produced in the market economies.

Other Minerals

Chromium mining capacity within the market economies is over 70 percent controlled by private mining corporations, operating mainly in South Africa and Rhodesia. The most important firms are General Mining and Finance Corp. Ltd. (South Africa), Union Carbide, Consolidated Mines Ltd., and Transvaal Consolidated.

Cobalt production in the market economies is dominated by majority government-owned enterprises in Zaire and Zambia. Cobalt is usually a byproduct or coproduct of copper.

Over 70 percent of the market economies' production of *molybdenum* is in the hands of private mining companies, including AMAX Inc., Duval, Placer Development, Molybdenum Corporation of America, and Kennecott Copper Corporation, almost all taking place in the United States.

To summarize this brief review, only in the cases of copper, iron ore and bauxite have expropriations during the last decade or so and the shift of ownership to governments been important to the structure of the industry as a whole. For nearly all the other nonfuel minerals, with the notable exception of cobalt, the bulk of the production—even in the developing countries—is by private firms. In some countries, such as Mexico and Malaysia, the restriction of foreign mining enterprises to a minority position tends to limit exploration and development.

The author offers the tentative conclusion that the threat of expropriation is no longer the dominant deterrent to foreign investment in the mineral industries of developing countries. Far more important today are the existing tax regimes and other factors adversely affecting such investment plus the fear that host governments will violate the provisions of contracts they make with foreign investors.

EFFECTS OF CHANGING OWNERSHIP PATTERNS

Mining industries do not necessarily become static after they are nationalized and begin to operate as government enterprises. In Chile, Peru and Zaire, the nationalized copper operations are being increased with the assistance of loans from both public and private international financial institutions. Brazil's state mining enterprise, CVRD, has expanded iron ore output aggressively, and a wholly Mexican-owned company (with minority government participation),

Mexicana de Cobre, is expected to complete construction of La Caridad by 1980, which at a rated capacity of 150 thousand tonnes (metric tons) per year will be one of the largest copper mines in the world.

These examples, and additional ones, indicate that it would be a mistake to assume that only multinational mining firms are capable of creating and expanding mining industries in the developing countries. It should be noted, however, that developing countries differ greatly in their level of industrial progress and in their capacity to mobilize the human, material and financial resources for the creation and expansion of a mining industry. Countries such as Chile, Brazil and Mexico have long traditions in this industry and contain within their borders the skills and managerial competence necessary to carry on operations. Many other developing countries cannot claim the same, of course. Thus, the special human and financial resources that international mining companies can bring to developing countries for the promotion of their minerals industries deserve careful attention.

In 1961, 52 percent of the copper mine-producing capacity of the market economies was in the developing countries; this percentage declined to 49 percent in 1965 and to 46 percent by 1970 (see Table II–1). In actual quantities, however, copper producing capacity in the developing countries grew by 1,142 thousand tonnes between the end of 1970 and 1976 as against an increase of only 507 thousand tonnes in the developed countries, so that in 1976 copper mine-producing capacity in the developing countries was 51 percent of the total for the market economies. A projection of capacity for 1981 based on planned additions, most of which are well under way, shows capacity in the developing countries rising by 826 thousand tonnes from the 1976 level to 4,475 thousand tonnes or 55 percent of total capacity of market economies by the end of 1981.[5] Most of the planned increases in copper producing capacity in the developing countries will be in Mexico, Chile, Peru, Zaire, the Philippines, and Iran.

This growth in copper capacity in the developing countries since 1970 provides an interesting perspective. Large increases in Chile and Papua New Guinea during the 1970s were actually initiated by foreign equity investments during the 1960s, and some major capacity expansions in other developing countries were financed by foreign direct investment during the 1970s in Peru (Cuajone) and Indonesia

---

5 The 55 percent projection in Table II-1 for the developing countries' share in total copper producing capacity of the market economies in 1981—made by the Controller's Department, Phelps Dodge Corporation— is slightly higher than the 52 percent estimate for the same year by the Council of Copper Exporting Countries. See *Outlook for Development in the World Copper Industry*, CIPEC (Paris, November 1977), p. 19.

TABLE II–1: COPPER MINE PRODUCING CAPACITY IN MARKET ECONOMIES, 1961, 1965, 1970, 1976, AND 1981 (PROJECTED)
(Vol. = 1,000s of tonnes)

	1961		1965		1970		1976*		Projected 1981	
	Vol.	%	Vol.	%	Vol.	%	Vol.	%	Vol.	%
Developed countries	1,934	48%	2,290	51%	2,988	54%	3,495	49%	3,657	45%
Developing countries	2,085	52	2,203	49	2,507	46	3,649	51	4,475	55
Total	4,019	100	4,493	100	5,495	100	7,144	100	8,132	100

* The 1976 capacity estimate differs from that given in Table C–2, partly because the latter includes the capacity of the Cuajone mine which began commercial production in 1977.

Source: Controller's Department, Phelps Dodge Corporation. See footnote 5 on page 21 regarding different projections for 1981.

(Ertsberg). As is discussed, however, foreign direct investment in the mining industries in the developing countries has generally declined during the 1970s. The vast bulk of the expansion in copper mining capacity in the developing countries that is scheduled to come into production by the end of 1981 has not been undertaken by foreign direct investment. Nevertheless, there are plans for foreign investment participation in the expansion of copper producing capacity in Chile, Zaire, Panama, Argentina, and Papua New Guinea, among others. However, this capacity, if realized, is unlikely to come into operation before the mid-1980s.

Bauxite and nickel producing capacity in the developing countries increased substantially during the 1970s, both absolutely and relative to capacity in the developed countries. This occurred notably in Botswana, the Dominican Republic, Guatemala, Indonesia, New Caledonia, and the Philippines in the case of nickel; and in Brazil, Guinea and Jamaica in the case of bauxite. Expansion of nickel producing capacity in developing countries resulted almost entirely from foreign direct investment, and, except for a substantial proportion of the Brazilian output, this was also true for bauxite. Iron ore producing capacity also rose substantially in the developing countries during the 1970s, but a large portion of the increase in Brazil occurred under the control of the government enterprise, CVRD.

From this discussion, it is clear that foreign ownership and investment need not be closely associated with the development of mineral producing capacity in developing countries—at least not in the case of all minerals and all countries. Although foreign investment undoubtedly supplies many benefits to the mineral industries of developing countries and in most cases is essential in the initial phases of these industries, it is not justified to assume that the development of mineral producing capacity in the developing countries is a simple function of foreign investment. Although foreign investment in mining in the developing countries is likely to become less of a factor in the development of known deposits, foreign exploration for new deposits is likely to continue to play a key role in future mineral development.

DECLINING FOREIGN INVESTMENT IN DEVELOPING COUNTRIES

While global figures on foreign investment in mining have not been compiled, there is considerable evidence that exploration and development expenditures in the developing countries by international mining firms have declined during the 1970s. Articles and speeches by U.S. and European mining officials, company annual reports and personal interviews by the author indicate that international mining firms are concentrating their exploration and development expenditures in the developed countries, especially the United States, Canada and Australia.[6]

U.S. Investment

Quantitative indications of declining U.S. investment in the mining industries of the developing countries during recent years are provided by two sets of Department of Commerce data. Between 1966 and 1972, the total book value of U.S. direct investment in mining and smelting in developing countries rose from $1,655 million to $2,267 million in current dollars (see Table II–2), but, five years later, at the end of 1977, direct investment stood at virtually the same figure following losses in 1974 reflecting the nationalization of U.S. mining investments in Chile, Peru, Zambia, and elsewhere. Compared to the equivalent values of U.S. mining and smelting investment in the developed countries (almost all in Canada and Australia),[7] the developing country share hovered at about 40 percent of the total during the 1966–72 period, but then dropped to 31 percent in 1974 and stood at 32 percent at the end of 1977.

This picture is substantiated by data for capital expenditures for mining and smelting of majority-owned affiliates of U.S. companies (see Table II–3). In current dollars, capital expenditures rose from an annual average of $244 million in the developing countries during 1965–69 to $292 million in 1970–74, but fell back to $227 million during 1975–78, and to far lower levels for the last two years. Measured in constant 1972 dollars, the annual value of this investment declined to about one-half in real terms between the period 1965–69 and 1975–78, and to one-third during the last two recorded years. If the large capital expenditures by the Southern Peru Copper Company (SPCC) in the Cuajone mine in Peru over the 1973–76 period are excepted, this decline would have been even more marked.

By contrast, U.S. plants and equipment expenditures in the industrialized countries averaged 27 percent higher in current dollars

---

6 See Thomas N. Walthier, "The Shrinking World of Exploration," *Mining Engineering* (April/May 1976); see also J.S. Carman, "A Case for Greater Investment of the United Nations in Third World Mineral Developments," paper presented at the 1977 AIME Annual Meeting, Atlanta, Georgia, March 1977 (mimeo); and *Mineral Development in the Eighties: Prospects and Problems*, a report prepared by a Group of Committee Members (British-North American Committee, November 1977), pp. 14–16.

7 At the end of 1977, the book value of U.S. mining and smelting investment in Canada and Australia accounted for 67 percent and 26 percent, respectively, of that in all foreign developed countries.

during 1975–78 than in 1965–69. It may also be noted that expenditures by U.S. firms at home for new plant and equipment in the mining industry rose from an average annual level of $1.8 billion over the 1967–72 period to $2.3 billion (in 1967 dollars) over the 1973–77 period.[8] These figures suggest that U.S. mining firms, including petroleum firms that have gone into mining, have been shifting their investment activities from foreign countries to the United States.[9]

Investment in Exploration

Because of the importance of exploration for the discovery and future development of the mineral reserves of developing countries, a decline there in these activities by the international mining firms, similar to that indicated for investment generally, may prove to be a highly significant and possibly ominous trend for long-run minerals supply. According to a survey of 18 U.S. and Canadian mining companies, over 80 percent of their exploration expenditures (including uranium exploration) in recent years has been in the developed countries.[10] Meanwhile, a study of the investment experience of 14 European mining firms revealed that between 1961 and 1975 the share of these companies' exploration expenditures in the developing countries declined from 57 to 15 percent,[11] and to 11 percent in 1976, followed by a recovery to 19 percent in 1977, due almost entirely to a large increase in exploration in Brazil.

The level of exploration expenditures in developing countries has been sustained to an increasing degree by the search for uranium, particularly in 1977. It is especially noteworthy that total expenditures by the 14 European mining firms in both the developed and developing countries have hardly increased at all for nonuranium

---

8 *Economic Report of the President* (Washington, D.C.: Council of Economic Advisors, January 1977), p. 247; and *Survey of Current Business* (March 1978).

9 Data on expenditures for investments in mining and smelting include exploration and development expenditures for uranium, which have been running fairly high. To this extent, data on investment expenditures overstate those for nonfuel minerals.

10 E.A. Scholz and A.J. Spat, "The Economics of Foreign Versus Domestic Mineral Exploration," *Mining Engineering* (June 1977); and M. Chender, "Copper Exploration Restrained by Resource Nationalism and Low Prices," *Engineering and Mining Journal* (August 1977).

11 European Group of Mining Companies, "Raw Materials and Political Risk," a report submitted to the President of the Commission of European Communities, 1976; see also "Need for Community Action to Encourage European Investment in Developing Countries and Guidelines for Such Action" (Brussels: Commission of the European Communities, January 30, 1978), p. 4.

TABLE II-2: BOOK VALUE OF U.S. DIRECT INVESTMENT IN MINING AND SMELTING, 1966-77
(Millions of current dollars)

	Developing Countries		Developed Countries		
	Value ($ millions)	Change from Previous Year (percent)	Value ($ millions)	Change from Previous Year (percent)	Developing Countries as % of Total
1966	$1,655	*	$2,328	*	42%
1967	1,791	+ 8.2%	2,658	+ 14.2%	40
1968	1,962	9.5	2,875	8.2	41
1969	1,970	0.4	3,029	5.4	39
1970	2,119	7.6	3,286	8.5	39
1971	2,218	4.7	3,569	8.6	38
1972	2,267	2.2	3,400	– 4.7	40
1973	2,265	– 0.1	3,773	+ 11.0	38
1974	1,784	– 21.2	4,007	6.2	31
1975	2,150	+ 20.5	4,398	9.8	33
1976	2,309	7.4	4,750	8.0	33
1977	2,265	– 1.9	4,802	1.1	32

* No entries here since new series began with 1966.

Sources: 1966 through 1975: *Selected Data on U.S. Direct Investment Abroad, 1966-76*, Bureau of Economic Analysis, U.S. Department of Commerce (1977). 1976 and 1977: U.S. Department of Commerce, *Survey of Current Business*.

TABLE II-3: ESTIMATES OF PROPERTY, PLANT AND EQUIPMENT EXPENDITURES IN MINING AND SMELTING BY MAJORITY-OWNED AFFILIATES OF U.S. COMPANIES: 1965-69; 1970-74; AND 1975-78 AVERAGES AND 1975-78 ANNUALLY
(Millions of current and 1972 dollars)

	Millions of Current Dollars			Millions of 1972 Dollars			Developing Countries as % of Total
	Developing Countries	Developed Countries	Total	Developing Countries	Developed Countries	Total	
1965–69 average	$244	$482	$ 726	$305	$603	$ 909	34%
1970–74 average	292	916	1,208	287	900	1,187	24
1975–78 average	227	613	840	164	442	606	27
1975	366	809	1,175	288	636	924	31
1976	262	672	934	196	502	698	28
1977	126	502	628	89	355	444	20
1978	155	470	625	102	309	411	25

Sources: U.S. Department of Commerce, *Survey of Current Business* (September 1974), pp. 25–31; September 1975, p. 34; September 1976, p. 24; September 1977, p. 26; and March 1979, p. 35. The GNP deflator used to convert current to 1972 dollars is given in the *Economic Report of the President* (January 1979), p. 186.

projects since 1972, which suggests a substantial decline in such exploration activities in real terms. A recent World Bank report indicates that as much as 80 percent of all exploration expenditures during 1970 to 1973 were in developed countries—particularly the U.S. and Canada.[12] As mentioned above, most of the ore bodies that have been developed or are being developed by national mining firms were originally explored by international mining firms. For example, Chile and Peru have inherited a number of large copper ore bodies on which considerable work had been done by international mining companies in the past, and it will be many years before all of these known ore bodies are developed.

The exploration that has taken place in the developing countries has tended to be concentrated in a relatively small group of nations that possess known mineral potential and have access to funds for exploration either from international corporations or from other sources. Exploration activity is especially heavy in Brazil, Chile, Indonesia, and the Philippines.

Some evidence shows that mining companies have reduced their exploration budgets generally and not simply in the developing countries. For most mining companies, profits have declined substantially since 1974, and exploration expenditures are likely to be cut back in periods when profits are low, or when companies are experiencing losses. In addition, many companies have proven large volumes of reserves, sufficient to meet their expected requirements for expansion of output for many years ahead. Given the high cost of exploration and the present uncertainties regarding demand for minerals, these companies are reluctant to increase outlays for "grass-roots" exploration. The period of highest cost exploration occurs at the time the data for a feasibility study must be assembled, but intensive exploration of a known ore body is generally not undertaken until a tentative decision is made to develop it.

Another deterrent to grass-roots exploration in developing countries by international mining companies is an inability to retain rights to future development of explored areas until conditions warrant exploitation, in other words, to bank reserves for future use. This may be because of minimum expenditure requirements or because of high rentals that must be paid to hold the areas. Finally, in the United States, uncertainties arising from environmental regula-

12 Mentioned in *International Finance*, Annual Report, National Advisory Council on International Monetary and Fiscal Policies (Washington, D.C., April 1978), p. 79. H. Brownrigg offers the following estimate for the developing countries' share of exploration expenditures in the mining industry—1961–65: 35 percent; 1966–70: 30 percent; 1971–75: 14 percent; in "Stabilizing the Political Risk Environment in the International Mining Industry" (London Business School, 1977).

tions that may inhibit the ability of mining firms to develop newly discovered reserves may have reduced exploration.

Although there may be enough identified copper ore bodies to meet the world's copper requirements without additional exploration for several years, this is clearly not true of tin and of a number of other metals. Moreover, it is always possible that new lower-cost sources of minerals might be discovered which could be expected to be exploited before the known lower-grade deposits are developed. However, this sequence implies continuous exploration. The lead time between discoveries from grass-roots exploration and the final development of reserves often runs into decades, and mineral producing countries must have a large inventory of ore bodies to enable mining industries to expand readily in response to changing market conditions for different metals.

Since 1974, the low prices of copper, lead and zinc have been another factor in the decline in exploration activity. Exploration for the fuel minerals—petroleum, coal and uranium—has increased substantially, of course, but again this activity has been highly concentrated in the developed countries, especially Canada, Australia and the United States.

Conclusion

It is quite possible that the decline in exploration in the developing countries may not have its impact for another decade or two on their ability to expand capacity in minerals development. To a considerable degree, they are living off the fruits of exploration undertaken by the international mining companies before the shift in ownership patterns described in this chapter. Eventually, the failure to maintain a warranted level of exploration will have its depressing impact on the growth of a mining industry. This, indeed, may be the most significant consequence of the declining interest of international mining companies in exploration in the developing world.

III. Majority-Owned Foreign Mineral Investment and Government Mining Enterprises Compared

A comparison of investment decisions, operations and financing between foreign investor-owned enterprises and government mining enterprises is essential for an analysis of two of the major concerns which this study is addressing.

(1) Will the shift in ownership and control to government mining enterprises limit the availability of minerals?
(2) Will government ownership lead to world overcapacity and thereby impair the mining industries of the developed countries?

As discussed at the beginning of Chapter II, these seemingly opposing concerns may not be wholly inconsistent. Long-term development of the world's nonfuel minerals industries may be impaired by constraints on the exploration and development activities of international mining companies. On the other hand, government investment decisions with respect to the expansion of production capacities of certain minerals may at certain times be at variance with the expected growth of demand in the medium term. Also, government mining enterprises have often been less than willing to adjust output to falling price and demand conditions. In this sense, the problem of surpluses generated by government mining enterprises is a short-run problem caused by the unwillingness of government enterprises to cut back on production and exports in the face of reduced world demand.

COMPARISONS OF THE TWO FORMS OF OWNERSHIP

Majority-Owned Foreign Private Investment

The Investment Decision-Making Process

A key question is whether the entry and even the dominance of government enterprise in the minerals industries of developing countries will alter the long-run relationship between mineral prices and total costs of production, including an adequate return on risk capital.[1]

---

1 Orris Herfindahl raised the basic question whether aggregate investment in the copper industry responds systematically to expected returns or whether investment is haphazard because

(Continued)

Although space does not permit a comprehensive treatment of the theory of investment in the minerals industries, we will review briefly a few essentials of the characteristics of investment decision making in international mining firms. It is perhaps useful to distinguish between aggregate investment policy of a mining firm and investment criteria for individual projects. The overall capital budget, including exploration outlays, of a large integrated mining firm is determined in part by the occurrence of investment opportunities which may arise from exploration activities or from acquisitions, in part by the availability of capital and in part by broad company strategies such as long-run growth, balance among stages of production in vertical integration, maintenance of the firm's world market share, the desire for diversification among resource investments, and company policies toward foreign investment, including its perception of political risk. Although investment decisions are influenced by a large number of factors reflecting a company's overall investment strategy, individual projects must meet certain minimum criteria of profitability.[2]

Decisions on sizable expenditures, whether at the exploration stage or for the construction of a mine or smelter, tend to be made only after the calculation of potential profits and of the probabilities of success. Large-scale exploration programs not only spread the risk by undertaking a number of investigations simultaneously, but the expenditures can be programmed with a view to achieving a targeted average rate of return on expenditures on a group of projects having varying probabilities of success.[3]

(Footnote 1 continued)
of accidental finds and the inability to predict the relationship between investment and returns. He concluded that investment in the copper industry does respond systematically to expected profit signals and that, therefore, it is possible to develop a long-run supply function relating output or capacity to costs of production and to copper prices. Herfindahl's study published in 1959 dealt with a period when there was relatively little government ownership in the mining industries of the market economies. Orris C. Herfindahl, *Copper Costs and Prices: 1870–1957* (Baltimore: Johns Hopkins University Press for Resources for the Future, 1959), Chapter 3.

2 This is true generally in a competitive industry such as the minerals industry. There are, of course, examples of companies with a monopoly position in an industry deliberately undertaking unprofitable investments in order to keep out potential competitors or deliberately creating excess capacity to discourage new entrants. Although this may have been the strategy in the past for some companies in certain branches of the metals industry, the change in the worldwide competitive structure of the minerals industries makes it unlikely to be a significant factor today.

3 For a more detailed discussion of the decision-making process in the exploration phase and in the mine feasibility study and of cash flow analysis and methods of calculating rates of return on mining projects, see Raymond F. Mikesell, *The World Copper Industry: Structure and Economic Analysis* (Baltimore: Johns Hopkins University Press for Resources for the Future, 1979), Chapter 8.

Before a mine is constructed, sufficient drilling, tunneling, sampling, and metallurgical testing of the ore body must be carried on to permit a feasibility study, and this is perhaps the most expensive aspect of exploration; it requires high-risk capital since there can be no assurance that the project will go forward until a full feasibility study is completed. This study not only projects the output of a mine in terms of expected metal content over a period of years and the expected costs of production, but also the size of the operation needed for economic feasibility and to recover the capital costs. From the information generated by the feasibility study, the revenues available for debt retirement and the rates of return on equity may be calculated within certain confidence limits. Minimum target rates of return on total capital at risk are usually expressed as the internal rate of return or discounted cash flow (DCF), or some variant of DCF. The feasibility study must take into account taxes and projections of various cost elements as well as projections of product prices. Unless the feasibility study shows a high probability that revenues after taxes will be sufficient to cover debt service by a reasonable margin plus a minimum rate of return to the equity holders, the mine will not be developed. The target DCF rate will reflect the opportunity cost of capital in other investments of a similar nature plus an allowance for the probabilities of exploration success and for political risk, especially if the investment is made in a foreign country.[4]

Projections of costs and prices must be made over a span of many years into the future. This is one of the reasons that mining investments are very hazardous, especially under current conditions of almost universal inflation which creates serious disparities between costs and selling prices. If the debt financing calls for a variable interest rate, even the debt service payments cannot be projected with any degree of certainty. Political risks arising from possible contract violations by the government of the country in which investment is made must be added to the commercial risks. Given all this, it is sometimes surprising that mineral investments are made at all.

When a large firm is making a number of investments, it should be expected that exceptional returns on some investments would offset lower than expected returns on others. Over a period of a decade or so of investments in various countries, and perhaps in different minerals, an *average* rate of return sufficient to cover the full costs of the operations, including returns to equity adequate to attract in-

---

4 Even in the United States or Canada, investments are subject to political risk arising from unforeseen government environmental regulations or changes in the tax regime, such as increased social security taxes. In Canada, the takeovers of resource investments by a Canadian province must be regarded as a significant political risk.

vestors, must be expected. This can never be the case for the large mineral development company if each government insists on changing individual contract terms whenever exceptionally high rates of return are realized on an investment in its country. Governments also have a tendency to look at accounting rates of return on book value during any particular year rather than at the DCF rate over an extended period of time. The DCF rate takes into account both the long period prior to the initiation of commercial production (or perhaps prior to the retirement of external debt), during which time there are no returns to the equity investor, and the variability of returns from year to year. There are, of course, ways of dealing with this problem in the negotiation of contracts, a subject which will be dealt with in the next chapter.

A further point to be made about the investment decisions of large private firms is their desire for permanence and growth of an investment. Corporations are not interested in risking their capital and employing their human resources over a period of many years to establish a successful industry and then fading away; this is true even if they are able to obtain a reasonable return on their investment and adequate compensation for their equity in the enterprise. Not only do they want permanent investments, they want to reinvest a share of the earnings to expand their investment if opportunities arise. Nor are mining firms usually interested in simply selling their technology and hiring out their experienced personnel (which constitutes their most important resource). This is the province of engineering and construction firms which generally take no risks and are organized for this type of business. See the discussion later in this chapter on the "depackaging issue."

The Foreign Investor's Contribution during the Construction Phase

Following the completion of the feasibility study, the most important function of the foreign equity investor in a large mining project is the mobilization of the debt capital.[5]

Because of the high capital cost of modern mines and the high risk of equity investment in developing countries, a substantial proportion of total capital, sometimes as much as 75 percent, will usually take the form of debt financing. This is derived from a number of sources: from equipment suppliers whose credits may be guaranteed by government agencies in the exporting countries; from commercial

---

5 Mobilization of debt financing is less of a problem for most petroleum companies that undertake mining investments, owing to their large cash flow from oil.

bank consortia for equipment and to finance "pipeline" shipment of the minerals; from loans from future consumers; and from credits from public national or international lending agencies, such as the Export-Import Bank or the World Bank. Dozens of creditors may be involved. In the case of the Cuajone mine in Peru, developed by ASARCO, over $400 million in debt financing was provided by some 50 banks, equipment suppliers and other lending agencies, including the International Finance Corporation (IFC).

Usually a *creditor agreement* must be negotiated to which both the foreign investor and host government will be parties. This agreement defines the obligations of both the host government and the foreign investor to the potential creditors. The foreign investor frequently has difficulty in obtaining the consent of the host government to conditions demanded by the potential creditors. Important among these conditions may be the setting aside of the export proceeds of sales of the mine's minerals under a kind of escrow arrangement which gives the creditors prior right to these proceeds over the host government in meeting debt service. The obligation of the foreign investor to the creditors is usually limited to that of constructing the mine in accordance with certain agreed standards and of achieving a minimum level of output as a percentage of rated capacity by an agreed completion date. For periods of up to four or five years, this can amount to a major liability of the mine development company, sometimes running into hundreds of millions of dollars, virtually without recourse.[6] In some cases, the foreign investor will also provide some of the debt capital in the form of advances which will be subordinated (in priority of repayment) to senior indebtedness provided by the external creditors especially in the event of cost overruns.

The external creditors usually require that the foreign investor negotiate long-term contracts for the sale of sufficient output from

---

6 The earlier BNAC report, *Mineral Development in the Eighties: Prospects and Problems*, a report prepared by a Group of Committee Members (British-North American Committee, November 1977), p. 23, had the following to say about these completion guarantees: "The mineral developer in an involved and costly deal usually has the obligation to prove to the lender that the project will work. This kind of assurance is normally enshrined in completion guarantees, which shift much of the risk from the lender to the mineral developer until the project is proven to the lender's satisfaction to work successfully. Meanwhile, the developer implicitly bears the major part of the total financial burden of the project. Such completion guarantees thus require a high degree of control by the developer over both the construction and the launching of a new project. In addition, the lender usually wants a similar guarantee of effective workability of the project and salability of its product during the whole period of loan repayment which, in the case of major mineral projects, may be 10 to 20 years. That is why 'turnkey' projects on the lines of those in easily designed factories involving much smaller investment are rare in large-scale mineral projects."

the mine to meet the debt service during the period of debt repayment. The negotiation of such contracts may be difficult, especially in periods of market glut, even though deliveries will not begin for several years when the mine begins to produce. In some cases, the parent firm itself must agree to accept a certain amount of the output.

Most mining firms employ the services of construction and engineering firms for much of or perhaps all the tasks involved in the creation of a mining complex. Even the feasibility study may be made by an engineering firm. However, officials and experts of the mining firm investing equity capital in the mine tend to work closely with the outside engineering and construction firms throughout the development process.

Long before a new mining complex is completed, foreign investors must establish a training program for the employees of the mine, the nature of which depends upon the quality and experience of the local labor force. This training program continues into the operating period because of labor turnover and, more importantly, in order to train local workers to take over the jobs of expatriates as rapidly as possible without sacrificing efficiency. After a decade or two, nearly all the skilled professional and supervisory personnel may be nationals of the host country, leaving only a handful of top managers from the parent company. In some cases, even the managers are replaced with nationals who become, in effect, a part of the organization of the international mining firm.

The Foreign Investor's Contribution during the Operating Period

Perhaps the most important single operating function of the foreign investor after start-up is professional management. It is often difficult for host governments to understand the unique contribution of management by the foreign investor once the mine is operating. They ask why this function cannot be assumed just as readily by the government itself, either by using its own nationals who may have considerable experience in the mining industry, or by hiring managers from abroad.

Several dimensions of management go beyond the fact that the host government may be able to find quite competent managers and that international mining firms do not have a monopoly on managerial competence. One dimension is political independence from the host government. Another dimension is the relationship between the manager appointed by the foreign investor and that person's international mining organization. Management of a large enterprise cannot be separated from policy decisions relating to budget allocations, planning, additional borrowing, arms-length negotiations with the host government, drawing on the resources of the parent company, and a host of other policy issues that are the province of the majority equity investor. Close liaison between the mining subsidiary and the parent company facilitates technical and financial problem-solving that is always important in new mine complexes during the early years of operation, as well as the utilization of the experience and R&D facilities of the parent company. If the host government were to hire the most competent, experienced manager in the world, that person could not maintain the relationship that the typical manager of a foreign subsidiary has with the parent company as a member of the policy-making team of that company. Nor could engineering and construction companies or mine management advisory companies substitute for an experienced international mining company with a majority equity interest in a subsidiary. The advantages of this parent-subsidiary relationship are often intangible and can only be described by actual case studies. This relationship provides a continual source of inputs from the parent company throughout the operating period, which cannot be duplicated by hiring outside services.

Government Enterprise

Comparisons between investment decisions and operations of majority-owned foreign investors in the mining industry, on the one hand, and those of government mining enterprises, on the other, are difficult to make for several reasons.

First, in most cases the mining industries in developing countries were initially established by foreign investors, and government enterprises have taken over long-established operations in which their own nationals had played important roles at every level up to and including top management.

Second, there are great differences from country to country in the competence and experience of government mining enterprises and in the number and quality of nationals with skills and managerial experience in the mining industry. For example, Chile has a long tradition in mining and was a major copper and nitrate producer and exporter for a century before the American and British mining firms came on the scene. Chile's government mining enterprise, CODELCO, is larger in terms of assets and trained and experienced mining personnel than many international mining companies. Much the same can be said regarding Brazil's government mining enterprises; and there are also large, well-managed national enterprises in Mexico. Conditions in these countries cannot be compared with Papua New Guinea, or even with Zaire and Zambia where large government-owned mining enterprises depend heavily for their operations on expatriates. Finally, examples of the same decision-making process and behavioral patterns and policies are to be found in some of the state mining enterprises as in the international mining companies.

Some of the advantages of close relations between national enterprises and international mining firms, including the flow of new technology, have been achieved by joint ventures involving a minority foreign investor with majority government investments. This is evidently the case with Anaconda's minority position in Minera de Cananea in Mexico, which Anaconda originally founded.

Therefore, broad generalizations regarding the behavior patterns of national enterprises in developing countries, whether government-owned or of mixed ownership, cannot be sustained by the facts. Success stories and dismal failures may be found in enterprises run by both foreign direct investors and by governments, and there are cases where government mining enterprises appear to be guided by the same profit-maximizing criteria as many international mining firms. Nevertheless, the two categories of enterprises do operate in somewhat different environments and are subject to different objectives and constraints. It is with these general caveats in mind that we look at the decisions and operations of state mining enterprises in developing countries.

Investment Decisions

It is sometimes said that government mining enterprises make investment decisions without proper regard for return on the investment, and are mainly concerned with employment, growth of the public sector, and net export earnings. It is true that governments do not have the same objectives as private enterprise and that political interests often take precedence over productive efficiency. Nevertheless, most government mine managers want to show that revenues cover costs as a measure of their success, and responsible governments are concerned with the allocation of capital on the basis of social opportunity costs. Moreover, if they are going to finance a mining venture with private or public international loans, government enterprises are expected to show on the basis of feasibility studies that expected returns from mining ventures are commensurate with those in other sectors of the economy. Certainly, internal rate of return criteria are stressed in project appraisals by private international financial institutions when considering loans for resource investment.

On the other hand, some costs that have to be borne by a foreign investor may not be included in cost calculations for a government mining enterprise in the same industry. Government enterprises are usually subject to taxation, but the burden of taxation to the government is usually not the same as that borne by private investors in the same industry. The government may also provide infrastructure in the form of roads, power, communities, facilities, and so forth, outside the capital budget of the government mining enterprise, whereas these facilities would need to be included in the capital budget of a private enterprise. Even where infrastructure costs might be included in the capital budget of a government mining enterprise, an allowance might well be made for benefits to the communities in the form of externalities, thereby reducing net social costs as contrasted with full monetary costs which would be borne by a private investor. Finally, the government of a developing country would not need to include in its minimum target rate of return an allowance for political risk as a foreign investor in that country normally would.

As just mentioned, cost calculations for a government enterprise may take into account social opportunity costs and social benefits rather than full monetary costs and money revenues, which would form the sole basis for project evaluation in the private sector. For example, if there were substantial unemployed labor, the social opportunity cost of employing the labor (who might otherwise be a burden on the state) would be less than the monetary cost. Likewise, externalities, such as the use by the community of highways, railroads, power facilities, and so on, may be considered a part of the social revenues from the government project, but these would not accrue to private investors. If the external value of the currency were believed to overvalued, export proceeds in foreign exchange might be valued at a shadow rate of exchange for purposes of social evaluation of the project in the public sector, although export earnings by the private investor would be valued at the actual rate of exchange.

Given a proper evaluation of the social benefits and costs, projects in the public sector would be assessed in the same way as they are in the private sector. However, the method of project evaluation employed by the government may lead to a government investment in circumstances where a foreign private investment would not be warranted. For example, governments frequently employ a rate of discount for evaluating public projects that is lower than the expected rate of return that would warrant a private investment in the same industry, although, as argued elsewhere, the rate of discount employed in evaluating public projects should be the same as that which is appropriate for the private sector in the same industry.[7]

Financing Government Mining Enterprises

It is sometimes argued that state mining enterprises in developing countries have lower financing costs because they can borrow from the World Bank or from the regional development lending institutions whereas foreign private investors cannot, except with a host government guarantee. It is for this reason that officials of private international mining firms have criticized the activities of institutions such as the World Bank and fear that an increase in World Bank lending to the extractive sectors of the developing countries may lead to world overproduction and constitute unfair competition by, in effect, subsidizing the expansion of mineral producing capacities in the developing countries.[8] This concern appears exaggerated, and, in any case, the problem should not occur if the officials of public international lending institutions employ proper loan criteria. The World Bank and the regional development lending institutions should not be expected to discriminate in favor of state enterprises in the developing countries as the charters of the public international lending institutions explicitly state that their lending is not to substitute for private international capital when available on reasonable terms. Moreover, the limited resources that the World Bank plans to allocate to the extractive industries in the developing countries and its declared policy indicate that its loans should serve as catalysts for expanding the flow of private equity and loan capital to the developing countries rather than be a substitute for such financing. See the discussion in Chapter V.

As regards private international lending, the point has already been made that subsidiaries of international mining firms are generally able to borrow from the international financial markets on better terms than government enterprises can obtain without a government guarantee. Government guarantees also impair the borrowing capacity of the government for other purposes.

Operating Costs and Efficiency

Mention has been made of the fact that government enterprise may be guided more by social opportunity costs than by monetary costs which must guide the operations of private firms. This probably

---

7 For an analysis of this issue, see Raymond F. Mikesell, *The Rate of Discount for Evaluating Public Projects* (Washington, D.C.: American Enterprise Institute, 1978).

8 See the American Mining Congress policy resolution in footnote 2, Chapter V.

explains the greater flexibility of private mining enterprises in reducing output in response to reductions in market prices. Government mining enterprises are under considerable pressure to maintain employment, and the failure to reduce variable costs may result in monetary losses. In countries where there is high severance pay, or restrictions on reducing the number of employees, private firms may also be constrained in reducing production. In some cases, this may in fact be uneconomical from a social point of view since there may be a shortage of experienced and skilled workers in the rest of the economy. Yet, even in developing countries, private enterprises probably have greater flexibility in shutting down or curtailing marginal operations than do state enterprises. State enterprises may also be under political pressure to locate additional facilities in politically important areas or in areas of high unemployment. State enterprises may also be required to purchase materials and services from state or other national sources at a higher cost than they could be obtained by open tender or from abroad. Political factors may even dictate the particular foreign sources of inputs for state mining enterprises. Whether the decision in any of these cases is "efficient" or "political" will depend to a great extent on the political independence of the managers of the state enterprises.

Conclusions on Government Mining Enterprises

A number of state mining enterprises are efficiently operated, but, although some governments have been able to expand their mineral industries, they have for the most part only continued to build on a foundation previously established by international mining companies, including the development of ore bodies originally explored by these companies. As a general rule, governments have not been successful in grass-roots exploration. The basic question, however, is not whether governments can operate and expand their minerals industries, but whether these industries can be developed and operated more efficiently in terms of capital and other resources, and as sources of tax revenues for the state, when foreign equity investment participates in developing, helping finance and managing the minerals project.

EVALUATION OF THE SPECIAL CONTRIBUTION OF FOREIGN INVESTMENT

How Important Is Foreign Direct Investment?

With production in some of the major minerals shifting to state ownership and control, and with less and less foreign private investment moving to mining and smelting in developing countries—very much less when measured in real terms—the question arises, how essential is foreign investment to the development of the world's mineral resources?

Officials of international mining firms usually regard foreign investment as virtually indispensable to the growth and efficient operation of minerals industries in developing countries. Many government officials, including trained and experienced mine managers in state enterprises, regard this attitude as chauvinistic, and point with pride to their own accomplishments. Moreover, many of them are convinced that foreign investment can be "depackaged" so that its various elements can be purchased at international market prices without paying "rents" to the international mining firms. Empirical support for both positions tends to be weak and slanted.

Some evidence shows that international mining firms can bring a new ore body into production more quickly and efficiently than can a government mining enterprise, and that reduced production and operational efficiency usually follow nationalization of a mining industry. For example, the Cerro Verde project was initiated by the Peruvian government enterprise, Mineroperu, at about the same time the Cuajone project was launched by the Southern Peru Copper Company. Cuajone came into operation at its full capacity of 155 thousand tonnes in 1976, but Cerro Verde did not come in until late 1978. Nevertheless, "best-worst" case comparisons do not provide an adequate basis for generalization, especially as some foreign private ventures have been disasters. More detailed case studies are needed to determine exactly which government mining enterprises have been relatively efficient and which have not, and for what reasons. If certain inputs are lacking or of poor quality, it is important to determine which, if any, of the specific elements in a foreign investment package cannot be transferred to the developing countries without a foreign equity commitment.

Without taking an extreme position on either side of the issue discussed above, the author suggests the following reasons why foreign investment is important for exploration and development of the minerals industry in developing countries.

(1) The larger the number of experienced international mining firms engaged in exploration and development in a country, whether it be Canada, Mexico, Peru, or the United States, the more efficient the mineral output of the country is likely to be. A government mining enterprise benefits from competition from private domestic and foreign mining enterprises that can provide

a yardstick for measuring government performance. Those who take the position that all mineral development, or the development of certain minerals, should be planned and implemented by government must answer the charge that government planned and controlled economies have had a far poorer performance record than free-enterprise economies, and that a government monopoly in the minerals area is not likely to perform any better than one in industry or agriculture.

(2) Foreign direct investment is more capable of providing politically independent management, whether the managers are nationals or foreigners. Top management and boards of directors of government enterprises not only tend to be political appointees, but are under constant pressure to put certain national economic and political objectives above that of running the mining enterprise with a view to maximizing gross profits. A major argument given for government operation of a mining industry is that this will assure its operation in the national interest. This relates to the efficiency issue raised in the previous paragraph: is it preferable to have a number of competitive enterprises each seeking to maximize net revenue within the framework of a general set of laws designed to protect the national welfare, or is it better to have a government planned economy?

(3) Foreign investment provides a continuing association with an experienced international mining firm for problem solving and for the introduction of new technology. Rarely do operations in mining and milling go as planned and without difficulties. No two ores are exactly alike, and problems in their treatment are very common in new mines. Following the expropriation of the large Chilean mines by the Allende government, technical consultants from all over the world were brought in both by Allende and later by the Military Junta to help solve production problems. Even so, output of the new installations which had been constructed prior to the Allende regime by Anaconda and Kennecott continued to be low and inefficient for several years.

(4) International mining companies are better able than governments to undertake the high risks of exploration, both because of the size of the financial resources they can command, relative to the budgets of many governments, and because of their ability to pool risks involved in many projects to produce a variety of minerals in many countries and regions throughout the world. Their ability to pool a large number of risks and to determine probabilities of success for each one reduces the overall cost of

risk-taking as contrasted with that for most national government mining enterprises.

(5) International mining companies are also better able than are government enterprises to mobilize the required technical and managerial inputs, together with the international loan capital, required for large mining projects. This is true even though government enterprises can hire the services of geologists, engineers, construction firms, and managers from a variety of sources throughout the world. Sometimes a government can hire an international mining firm to provide all the inputs for each phase of the development of a mine. This was the route taken by the government of Iran in the development of Sar Cheshmeh under a technical assistance contract with Anaconda.[9] Such arrangements are costly and expose the government to all of the risks. Moreover, international mining firms tend to employ their best technical and managerial resources for their own projects, rather than use them on contract projects.

(6) In the case of minerals such as bauxite, iron ore and manganese for which organized international markets and commodity exchanges do not exist, vertically integrated international mining firms can provide a market for the output. Even for commodities such as copper, much of which is traded on a worldwide competitive basis, the marketing organizations and affiliate relationships of international mining firms are generally more efficient than the marketing staffs of government mining enterprises in the developing countries. The negotiation of long-term contracts to take the bulk of the output of a new mine during the debt repayment period is frequently a condition for obtaining the necessary international loan capital for its construction. International mining firms are in a better position to negotiate such contracts.

(7) Equity participation by an international mining firm is frequently indispensable for raising the large amount of private or public international debt capital required for a modern mine. Many large mines are now costing one-half billion to one billion dollars, and most of this capital must be raised by borrowing from international banking consortia and by credits from foreign suppliers of equipment. Governments of developing

---

9 Anaconda was in an unusual and special position, having just lost its Chilean mines to nationalization. As a consequence, it had a large cadre of experienced mine managers and technicians, only recently out of work and ready for transfer to Iran without regard for usual financial and other arrangements.

countries usually prefer to have such borrowing done by the mining enterprise rather than obtaining the funds directly in the form of a public external debt transaction, since the latter would reduce the government's own borrowing capacity. International creditors usually look to future mineral exports of the mining project to service the debt. This means that the creditors will insist on the construction and management of the mine by an experienced international mining firm which has an equity stake in the venture. Unless the government enterprise has an exceptionally good reputation based on proven experience, public international development institutions such as the World Bank also require participation by an experienced international mining firm as a condition for financing the enterprise.

An indication of the importance of foreign investment in the minerals industry is the desire on the part of experienced national mining enterprises such as CODELCO in Chile, Mineroperu in Peru, CVRD in Brazil, Mindeco in Zambia, and Penoles (a large privately owned Mexican mining company) to have international mining companies as their partners in their mining ventures. Moreover, a number of countries that have nationalized large foreign mining enterprises in the past are welcoming majority-owned foreign companies to explore and develop mines to extract the same minerals produced by the nationalized mines. This is true, for example, in Chile and Zaire. If the nationalized mining companies mentioned above believe that they can acquire all of the inputs supplied by foreign investors without accepting foreign equity investment, why are they willing to share the "rents" from mineral exploitation with foreign firms? If the governments of many developing countries with their own state minerals enterprises believe that their national welfare is maximized only by government or national private exploitation of minerals, why do they welcome majority-owned foreign enterprise to share in the development of their minerals resources? The reasons given by government officials of these developing countries are the same as those outlined in the previous paragraphs: the need for high-risk capital, foreign technology and management, marketing outlets, and the mobilization of international loan capital.

The advantages of foreign investment listed above apply to both developed and developing countries and not only is the distinction between these two categories of countries somewhat arbitrary, but the flow of technology, skills and experience among the developed countries is just as important in the resource industries as it is now recognized to be in manufacturing.

What is being argued here is not the absolute superiority of any structural form in the mining industry of any country, developed or developing, but rather the desirability for foreign investors to have the opportunity to enter any market economy to exploit resources under appropriate regulations applicable to all mining enterprises and to adopt whatever organization form is most conducive to the international flow of investment, e.g., 100 percent foreign ownership or joint venture with national private or government enterprise.

This conclusion does not suggest that traditional 100 percent or even majority-owned foreign investment is necessarily the most efficient or the most desirable in all cases. It does suggest, however, that foreign equity investment has substantial advantages to the developing countries as a vehicle for the transfer of technology, skills and management as contrasted with simply purchasing or hiring these inputs from abroad—if it could be done.

The "Depackaging" or "Unbundling" Issue

All mining industries in developing countries, and many in developed countries as well, require a variety of resources, skills and other inputs which have to be obtained from foreign sources. These are especially important at the stage of exploration and development of new mines.

The unique contributions that can be made by foreign mining firms have been detailed in this chapter and it was pointed out that with few exceptions these contributions can only be made through equity participation by these firms. They can provide the risk capital for exploration, localize the debt and equity financing for the project, negotiate long-term contracts for the sale of the output, furnish experience and politically neutral management, and provide a channel for technical and financial problem solving and for the introduction of new technology.

Yet, there is a widespread view that these foreign contributions needed for the development and operation of mining industries in Third World countries can just as readily be acquired without foreign equity investment. Cannot all of these foreign inputs, it is asked, be hired by government enterprises, piece by piece, without such investment? Would not this save paying the "rents" which foreign investors expect to receive in the form of returns on their equity investment over and above the rates of interest on international capital? Specifically, is it not possible to hire basic technology for mines and processing plants from the same engineering and construction firms as the foreign investor would use—whether the mine has majority foreign equity ownership or state ownership?

This is the so-called depackaging or unbundling issue.

It is true that engineering and construction firms, such as Bechtel Corporation, Arthur C. McKee, Fluor-Utah, Kaiser Engineering, Parsons-Jurden, and H.A. Simons, among others, operate in virtually every phase of mine development. Their clients include all of the large international mining firms for which they perform a variety of services such as the preparation of feasibility studies and the construction of underground or open pit mines, concentrators and smelters. Nearly all of the large mines constructed in recent years, including the Bougainville mine in Papua New Guinea and the Caujone mine in Peru, have involved the services of one or more engineering and construction firms.

It is also the case that these same firms provide a variety of services for state mining enterprises. For example, Parsons-Jurden has prepared a feasibility study for Mineroperu's Santa Rosa mine and for the expansion of Centromin's Cobriza mine for which the state-owned enterprise obtained World Bank financing. Fluor-Utah of California has reportedly been commissioned to do a feasibility study for Centromin's Totmocho copper ore body, and the Canadian firm H.A. Simons carried out the feasibility study for Mineroperu's Tintaya copper ore body.

Yet, consulting and engineering firms do not take equity positions or provide direct financing, and only rarely do they provide managerial services during the operating period. These firms work for a fee and require advance payments to cover their fees and reimbursable expenses. They do not answer all the other needs for foreign inputs in developing countries.

Indeed, it is very doubtful that all of these inputs are available for hire separately. There is considerable evidence that they are not. Furthermore, the package of resources that is integrated within an experienced mining firm is far more effective than the sum of its parts.

There is a real question, moreover, whether states can adequately take—or should try to take—all the risks of exploration, mine development and marketing, as well as all the technological challenges. Let us assume that a state-owned enterprise was willing to take all of these risks, and could properly evaluate them as costs, and pay the "rental prices" for all the inputs it would need to hire in the absence of an equity interest by an international mining firm. It is still questionable that it could develop the nation's minerals resources as effectively from the standpoint of the national economic interest as could a competent foreign equity investor, although admittedly governments do not make decisions on grounds of economic effectiveness alone.

A government need not decide between public and foreign private enterprise for all of its resource development. It may operate a state mining enterprise in some areas or products while encouraging foreign and domestic private mining firms to explore and develop minerals resources in others. The government may also enter into joint ventures involving a foreign private investor and a state mining enterprise. One thing seems clear: the participation of foreign mining enterprises will greatly enhance the amount of exploration and the rate of development of its minerals resources as compared with the situation in which the government establishes a state monopoly of all exploration and development.

IV. Conflicts and Accommodations Between Foreign Investors and Host-Country Governments

Even where the government of a developing country with mineral potential was willing to accept foreign equity investment, there have been a number of obstacles to the flow of such investment in recent years. For example—

(1) The general investment climate was often characterized by economic and political instability.
(2) There has been, in addition, a long history of either expropriation or broken agreements with foreign investors. This story is too familiar to bear repeating. In addition to violating specific contractual provisions, countries have demanded renegotiation of contracts on terms more favorable to the government; they have introduced new taxes, including export taxes, production taxes or taxes on reserves; or they have restricted transfer of dividends and the repatriation of investment capital; often they have enacted labor and social legislation that substantially increased labor costs, including requiring the employment of surplus labor; sometimes they have nationalized or otherwise interfered with the marketing of the product.
(3) There have been frequent conflicts between potential or existing foreign investors and host governments over the terms of investment agreements or of joint-venture agreements involving state enterprise as one of the parties.
(4) Sometimes the policies of the home government of the parent company adversely affect the investment of their nationals in foreign extractive industries; and there have been many uncertainties with respect to the future of such policies in the years ahead.

This chapter is principally concerned with the third of these obstacles—conflicts over the terms of *investment agreements*—and also with approaches to the resolution of these conflicts. These agreements are usually negotiated within a legislative framework, but frequently the legislation gives a foreign investment committee or other government regulating agency rather broad powers, including the power to negotiate contracts that provide for an exemption from existing laws, or that guarantee an exemption from laws that may be enacted in the future. Frequently, a foreign investment agreement is required as a supplement to a joint-venture contract with a government enterprise. For example, in Chile where two joint ventures involving a government mining enterprise and a foreign company were negotiated with relative ease, the negotiation by the two foreign companies with the Chilean Foreign Investment Committee required two years.

Mine development agreements have become quite comprehensive and cover many aspects of mine operations, including exploration, construction, financing, distribution of earnings, employment, advancement of locals in management, imports, marketing, foreign-exchange transactions, taxation, relations with the community, environmental controls, and safeguards requested by the foreign investor against changes in contract provisions. Negotiations often take many months, in part because of the detailed nature of the contracts, and in part because of the time required to reach compromises on opposing positions. Modern negotiations tend to be quite sophisticated, with each side armed with specialists in geology, mining engineering, financial accounting, taxation, environment, and other fields relating to social and economic impacts. Proposals and counterproposals are tested against their implications for the company's discounted cash flow, or its ability to mobilize debt financing, on the one hand, and against the fiscal implications for the government, on the other. Each side may now use computer technology for analyzing the financial impacts of proposals. Frequently, the mining agreement must also take into account the requirements of the creditors who will be expected to provide the debt financing for the project. In some cases, representatives of the potential creditors, including the World Bank, are brought into the negotiations. Their demands will often strengthen the position of the foreign investor in the negotiation.

The usual bargaining strategies are employed, including threats by one side or the other to break off negotiations. However, there is a strong impression that the outcome of such a negotiation is determined more by the willingness of each side to seek compromises that will achieve the minimum requirements of both sides, the company and the government, than by bargaining strategies. In a major negotiation, each side is too well informed for clever bargaining strategies to play a critical role.

CONFLICTING DEMANDS

Conditions Generally Desired by the Foreign Investor

The "ideal" contract conditions generally desired by the foreign investor are:

(1) majority equity ownership;
(2) full control over production, employment, investment, purchases of materials and equipment, marketing and distribution of earnings;
(3) tax provisions that will enable the foreign investor to earn and repatriate the capitalized value of the investment, including the repayment of external indebtedness, within a relatively short period of time, and a corporate tax rate that does not exceed that imposed by the home country of the parent company;
(4) foreign-exchange arrangements that permit the foreign investor to hold sufficient export proceeds abroad to meet all external obligations, including those arising from current foreign purchases, and to remit dividends and authorized capital repatriation;
(5) freedom to make payments to the parent company, or to other foreign firms, for technical services and the use of patented processes and to import equipment and materials from any source so long as the prices and quality are competitive;
(6) exemption from import duties on equipment and materials employed in construction and operations;
(7) no export or production taxes or royalties, and guarantees against the imposition of new taxes or other legislation or regulations that would affect the operation and profitability of the investment that did not exist at the time of the investment agreement, or were not specified in it;
(8) guarantees against expropriation or contract renegotiation during the life of the agreement;
(9) negotiation of an agreement covering exploration, production and operations before any substantial exploration outlays;
(10) a minimum tenure of the life of the mine, often at least 30 years after the initiation of commercial production;
(11) in the event of expropriation, a guarantee of full compensation based on replacement cost of assets or present value of projected earnings;
(12) arbitration of disputes arising in the implementation of the contract by an independent arbitration agency such as the International Chamber of Commerce Court of Arbitration or the World Bank International Center for the Settlement of Investment Disputes.

This list could be extended, but it probably includes the most important initial or "ideal" demands of the foreign investor.

Conditions Demanded by Host Governments

The governments of host countries also usually insist on certain "ideal" conditions from their point of view for foreign investments in their resource industries. Although these demands differ from country to country and from time to time, they frequently include the following:

(1) majority government ownership, or the option to acquire a majority of the equity at some time following the exploration period; and, in some cases, the option to acquire 100 percent of the equity after a certain number of years of operation with compensation to be made at book value and often in long-term bonds;
(2) gradual replacement of all expatriate personnel by nationals, and the establishment of training programs designed to achieve localization targets in accordance with rigid timetables;
(3) high excess-profits tax on accounting profits in any year resulting from higher than anticipated market prices, with no carry forward and no accelerated depreciation;
(4) repatriation of all export earnings to the central bank and the application of existing foreign-exchange regulations to the foreign investor;
(5) government control over marketing of the products;
(6) domestic processing of minerals and gradual expansion to downstream operations;
(7) the right to demand contract renegotiation whenever the government decides that changes in the contract are in the national interest;
(8) full application of all national laws to foreign investors with no guarantee against changes in the legislative framework in contracts with foreign investors;
(9) the settlement of all disputes between the government and the foreign investor through national judicial procedures and the rejection of any form of international arbitration.

Approaches to Reconciliation of Positions

Some or all of the government policies and conditions set forth above are likely to be incompatible with those regarded as essential by foreign investors. Without compromises, new investments in resource industries are unlikely to be made.

In recent years, there have been some new and imaginative contractual arrangements that go a considerable distance toward meeting certain of the policy objectives of host governments while at the same time satisfying minimum requirements of the foreign investor. Thus, compromises, short of either side's "ideal" positions, are being made in the negotiation of major new mineral investment agreements.

Some notable examples of reconciliation found in recent mine development agreements are:

(1) delegating management control by means of a long-term management contract to a foreign investor having only a minority equity participation. In addition, the minority investor may be given an equal vote on the board of directors of the company with respect to certain important policy issues, such as the distribution of earnings and investment and financing policies;
(2) sharing high-risk exploration outlays by the government and the foreign investor in proportion to their equity ownership in the mining enterprise, with full payment by the government for its equity share in the enterprise;
(3) negotiation of a mine development contract that covers all phases of the development of a mine, including exploration, feasibility study, mine construction, and mine operation, but providing for the maturity of the contract in, say, 30 years after the beginning of commercial operations;
(4) establishing a tax formula in the contract which ensures full repatriation of initial capital before any corporate tax is levied on earnings. This can be accomplished through accelerated depreciation or by means of a tax holiday. In addition, the tax formula may provide for carry-forward provisions that ensure no more than an average agreed tax rate on earnings over a period of years. Finally, the tax formula may provide that no more than the normal tax rate would be levied until the investor has earned an agreed minimum rate of discounted cash flow on the investment, after which excess-profits tax rates would apply. The excess-profits tax could be eliminated in years when the accumulated earnings reflect a discounted cash flow at a rate lower than the agreed minimum;
(5) depositing foreign-exchange earnings in a special account of the host's central bank at a foreign commercial bank under an escrow or trustee arrangement whereby debt service and other agreed obligations must be paid from this account before funds become available to the government;

(6) exempting the foreign investor from any new taxes, such as production taxes, export taxes, import duties, and so forth, for a period of years following the initiation of commercial operations;
(7) providing for a review of contract provisions a certain number of years after the full repatriation of the investor's capital, together with safeguards against demands for contract revision that would significantly impair the profitability of the investment;
(8) giving the government the option to acquire all or a portion of the equity shares after a certain period, perhaps 20 years, of commercial operations, with the terms of payment fully specified;
(9) giving the investor the option of financing with a high debt-equity ratio, with all or a portion of the debt constituting parent company advances, the interest and principal payments on which are not made subject to local taxation;
(10) providing for the arbitration of disputes over the interpretation or the implementation of the contract in ways satisfactory to both the foreign investor and the government.

Some version of each of the above approaches to conflict resolution may be found in the following summary of selected recent mine development agreements.

SUMMARIES OF FIVE MINE DEVELOPMENT AGREEMENTS EMBODYING INNOVATIVE CONTRACT PROVISIONS[1]

(1) The Papua New Guinea-Bougainville Copper Ltd. Agreement (1974)

The 1974-renegotiated agreement between the PNG government and Bougainville Copper Ltd. represents an attempt to deal with a possible future contract renegotiation in an orderly manner. It also

---

1 Not all of the recent mining agreements discussed in this section have been published, but the author has obtained copies of the texts of each of them.

In the analysis and interpretation of certain of these agreements, the author has benefited greatly from articles written by Steven A. Zorn, "New Developments in Third World Mining Agreements," and Thomas W. Walde, "Lifting the Veil from Transnational Mineral Contracts: A Review of Recent Literature," both in *Natural Resources Forum* (Vol. 1, No. 3), April 1977.

contains some tax provisions designed to protect the company from the effects of inflation on the tax rate.

During the first year and a half following the beginning of commercial operations of the Bougainville mine, profits were exceedingly high in relation to what they were expected to be at the time of the negotiation of the mining agreement in 1967. This was the result of higher than expected prices of both copper and gold, an important by-product. Although profits declined sharply with the fall in copper prices after mid-1974, the high initial profits led to a demand by the PNG government for a renegotiation of the contract. The 1974 agreement provided for a review and possible contract renegotiation every seven years. The language of the review provision suggests that changes in the agreement are to be made only on the basis of mutual consent. The arbitration arrangement, which provides for the appointment of a third arbitrator from a panel of five arbitrators to be nominated by the President of the Asian Development Bank, would appear to safeguard the company against being forced to accept highly unfavorable conditions at the time of this periodic review.

The 1974 agreement provided for a normal rate of corporate tax (currently 36.5 percent) on corporate income up to 15 percent of the capitalized value of the investment (adjusted for future additions to capital investment), and a 70 percent corporate tax rate on taxable income above this amount.[2] Since the capitalized value of the original investment does not change (except for capital replacements), a continuous high rate of world inflation might mean that a substantial portion of the taxable income would be subject to a 70 percent corporate tax rate, while the real value of this income in, say, 1974 prices may not have risen. However, provision is made for an adjustment of the tax formula in favor of the company under conditions of "abnormal inflation" (defined as an annual increase in the U.S. consumer price index in any tax year that exceeds by 20 percent or more the average rate of increase in that index in the five years ending with the tax year). Although the tax laws provide for payment of taxes in PNG currency, i.e., kinas, the tax formula provides for an adjustment with a change in the U.S. dollar value of the kina, and, in addition, provides for an adjustment in the tax rate in the event that the value of the U.S. dollar in terms of special drawing rights (SDRs) varies more than 10 percent from such value in November 1974.

2 For a discussion of the renegotiation of the PNG-Bougainville Copper Ltd. agreement (initially negotiated in 1967), see Raymond P. Mikesell, *Foreign Investment in Copper Mining* (Baltimore: Johns Hopkins University Press for Resources for the Future, Inc., 1975), pp. 130–132.

(2) The Broken Hill Proprietory Agreement on Ok Tedi (1976)

The Broken Hill Proprietory (BHP) agreement with the government of Papua New Guinea in 1976 for the exploration and development of the Ok Tedi copper mine illustrates the introduction of the discounted cash flow rate principle in tax arrangements. The negotiation of this agreement followed a long and ultimately unsuccessful negotiation with Kennecott Copper Corporation for the development of the Ok Tedi mine as well as the successful renegotiation of the Papua New Guinea government's Bougainville agreement with Conzinc Riotinto of Australia, from which the government and its foreign advisers learned a good deal. The BHP agreement was signed in March 1976, but did not become final until October 1976 when Amoco Minerals and a German consortium made up of Metallgesellschaft, Siemens, Degussa, and Kabell und Metallwerke joined the project. BHP and Amoco will each hold 30 percent of the equity, the German consortium 20 percent and the PNG government 20 percent (as it does in Bougainville Copper Ltd.).

The companies can claim accelerated depreciation sufficient to give a total cash flow in each year of at least 25 percent of the initial investment. Thus, if earnings are adequate, the project should earn its original investment back in four years. After recovery of initial capital, the normal corporate tax (36.5 percent) plus a dividend withholding tax of 15 percent will apply until the mine has earned a "reasonable return," which is, in effect, defined as a 20 percent discounted cash-flow return on total investment in the project (equity plus debt capital). For profits in excess of this rate of return, the corporate tax rate rises to an effective rate of about 58 percent. For example, with a debt-equity ratio of 3 to 2 and an average interest rate of 8 percent on debt financing, the foreign investor could earn a DCF return on equity of over 35 percent before the excess profits tax becomes effective, assuming no reinvested profits.

In this agreement, the equity share of the PNG government will be covered in part by the work that Kennecott had already done on the project, for which $17.5 million in bonds has been offered to Kennecott by PNG, subject to the actual construction of the Ok Tedi mine. Also, any infrastructure provided by the government, even if financed by loans from agencies such as the World Bank or the Asian Development Bank, will be credited to the government's equity interest.

The foreign companies, which hold 80 percent of the equity, must bear the entire risk of the project until completion of the feasibility study, which was expected to be completed at the end of three years at a cost of $12 million. If the foreign companies decide not to go ahead with the project after the feasibility study has been completed, they have no further obligation. However, they do bear all the risk up to this point. There is no provision for a management fee as the foreign companies will have full control of the management.

According to the BHP-PNG agreement, the company must prepare proposals to be approved by the government for the progressive replacement of foreign technicians, operators, supervisors, clerical, professional, administrative, and managerial staff, and for a training program designed to achieve this objective. The agreement also provides for the preparation of an environmental impact study, the scope of which is set forth in some detail, and the company is required to comply with specified environmental protection standards.

Arbitration procedures are also set forth in considerable detail. Each side nominates one arbitrator, and if they cannot agree on a third arbitrator, that person shall be appointed from a panel of five arbitrators to be nominated by the President and Chairman of the Board of Directors of the Asian Development Bank.

The agreement recognizes the right of the company to retain outside PNG the foreign exchange proceeds of exports to the extent necessary to enable the company to meet its foreign-exchange obligations or to pay dividends to overseas shareholders.

(3) The Texasgulf-Panama Agreement on Cerro Colorado (1973)

The Texasgulf-Panama agreement for the development of the Cerro Colorado copper ore body illustrates the use of the management contract and other contract arrangements for the protection of a minority foreign equity holder. The agreement also illustrates the provision for future government acquisition of the foreign investor's equity under terms and conditions specified in the agreement. Finally, this is the first investment agreement with a Latin American country in recent years that provides for external arbitration of disputes.

The Texasgulf-Panama agreement was preceded by lengthy and ultimately unsuccessful negotiations by the Panamanian government with two other firms, Canadian Javelin and Noranda.

In this agreement, Texasgulf holds only 20 percent of the equity, the remainder being held by the Panamanian government enterprise, CODEMIN. Under this agreement, the Panamanian government will share in proportion to its equity interest the cost of the feasibility study, and, in addition, is reportedly paying $20 million (in bonds) to Canadian Javelin for the work that it did on the project. Texasgulf's earnings from its 20 percent equity participation will be supplemented by a management fee (initially at a rate of 1.5 percent of gross sales, but declining to 0.75 percent) over the 15-year period of the management agreement, plus a sales and marketing fee. The joint venture will pay a 50 percent corporate tax to the Panamanian government, and there is a 10 percent dividend withholding tax.[3]

Despite the fact that Texasgulf will have only 20 percent of the equity in the project, management control is guaranteed to Texasgulf for the first 15 years of commercial production. After this period, management will revert to the Panamanian government. In addition, the Panamanian government has the option of acquiring Texasgulf's 20 percent equity at the end of 20 years at a price based on average earnings during the prior 5 years. It is interesting to note that during this 5-year period, management will be fully in Panamanian hands so that Texasgulf would be encouraged to do a good job in training the new managerial team.

The management agreement provides that in hiring personnel, preference shall be given to Panamanians in all job classifications. The administrators are to submit to the board of directors of Cerro Colorado S.A. (the operating company) a program for the training and instruction of personnel, which is designed to achieve the gradual transfer of all employment classifications to Panamanian personnel; such transfers are to be virtually complete at the expiration of the management agreement. This program, together with subsequent modifications, requires the approval of the board of directors of Cerro Colorado S.A. on which Panama will have a majority vote.

Disputes arising in connection with the management agreement are to be settled by arbitration under the Rules of Procedure of the Inter-American Commercial Arbitration Commission. If the two arbitrators chosen by the parties to the dispute are unable to agree upon a third arbitrator, the Commission designates the third arbitrator and the decisions of the arbitral tribunal shall be by simple majority. However, the contract provides that "judgments of execution of the arbitral awards" are to be issued by "courts of justice of the Republic of Panama." Such arbitral awards "shall be considered as if they had been rendered by Panamanian arbitral tribunals in accordance with provisions of laws presently enforced." Presumably,

---

3 The author was told by one of the Panamanian negotiators that the management, sales and marketing fees, together with the tax arrangements, were found to yield a 23 percent DCF rate of return to Texasgulf on its equity investment, and that DCF rate calculations were employed in the course of the negotiations. However, this was not confirmed by the Texasgulf officials whom the author interviewed.

this language was used in order to preserve the principle that all matters of litigation shall be in accordance with Panamanian law.[4]

(4) The Chilean Mining Agreements (1977)

The 1977 mining agreements negotiated between the government of Chile, on the one hand, and individual foreign mining companies—St. Joe Minerals Corp., Noranda, Falconbridge and Metallgesellschaft—on the other, illustrate the principle of guaranteeing the foreign investor a maximum tax burden while at the same time allowing that investor to be taxed at the domestic corporate tax rate if that rate is lower than the guaranteed rate.

Under the 1977 Chilean agreements, the total income tax burden on the foreign companies is fixed at 49.5 percent of net profits (including the housing tax and the dividend withholding tax) for the life of the agreement. However, should the normal tax regime applicable to domestic firms in Chile be reduced during the life of the contract, the companies have the right to elect the normal tax, but in this case they lose the guarantee that the tax applicable to them will not be changed. The foreign companies are also guaranteed exemption from any new taxes on production or exports.

The 1977 Chilean agreements give the foreign investors the right to retain abroad export proceeds of an amount sufficient to meet debt service and certain other obligations which become due within a stated period of time. The agreements also provide that the foreign investor will have the right to retain abroad foreign exchange in an amount equal to profits that have been delayed in remittance for more than one year after the date of application for transfer, in compliance with applicable laws and regulations.

The agreement with St. Joe Minerals Corp. gives the company the right to contribute up to 85 percent of the external capital in the form of advances, thus facilitating repatriation of capital without being subject to the corporate tax.

(5) The Indonesian-RTZ Agreement (1977)

The 1977 Indonesian agreement with Rio Tinto Zinc illustrates a method by which the host government's requirement for the repatria-

4 Whether this provision actually avoids compromising the "Calvo doctrine" on international arbitration is something on which the author is not qualified to express an opinion. However, one of the Panamanian negotiators told the author that if a dispute under the agreement ever goes to international arbitration "the agreement is dead."

tion of all export proceeds can be reconciled with the need by both the company and its creditors to retain foreign exchange from export proceeds to meet debt obligations and to transfer net profits.

The unwillingness of the Indonesian government to grant foreigners the right to retain export proceeds abroad needed to meet foreign currency obligations was a major barrier to the negotiation of any mine development agreements in Indonesia for about four years prior to the agreement with RTZ in March 1977. The issue was settled in the agreement by the adoption of an arrangement whereby export proceeds would be deposited in a foreign investment account held in a foreign bank in Indonesia in the name of the Bank of Indonesia as agent for the company. Such funds would be available for use by the company in discharging its obligations and for transferring net profits and depreciation on imported capital assets, according to regulations set forth in the mining agreement.

A similar arrangement was negotiated between Southern Peru Copper Company and the Central Bank of Peru for handling export proceeds of the Cuajone mine needed to meet SPCC's external debt obligations. In that case, however, the account is held in a New York bank in the name of the Central Bank of Peru.

Indonesian participation in the ownership of the operating subsidiary, PT Riotinto Indonesia, is made possible under the agreement by providing that RTZ will offer to sell shares in the company to the government or to Indonesian nationals in each year following the end of the first full calendar year after commencement of production. The offer of shares in each year is not to be less than 5 percent of the total number of shares outstanding when the offer is made, and the offer of such shares shall be made on terms and conditions reasonably intended to ensure that such shares are not thereafter transferred to non-Indonesians. The company must continue to offer shares each year until 51 percent of the shares in PT Riotinto Indonesia are held by private Indonesians or the Indonesian government.

CONCLUSIONS

Some of the recent mine development agreements seem to exemplify promising approaches to resolution of conflicting demands or preconditions of host governments and foreign investors. Admittedly, many of the provisions in these agreements would not be acceptable to some mining firms or to some governments, but they do offer a certain amount of hope that more imaginative agreements are one avenue for achieving larger participation of international mining firms in the developing countries.

Of course, the best agreements possible will mean little if their provisions are denounced by succeeding governments or if their provisions are modified unilaterally by the governments that make them. There are plenty of examples of such bad faith. On the other hand, a certain amount of optimism seems justified. For one thing, some of the disputes that have arisen out of the implementation of mining contracts in the past were definitely a reflection of poor drafting of the contracts rather than of outright bad faith on the part of the host governments—or of the companies. Also, most governments now seem to feel more self-confident in a way that they did not under the older style concession agreements. Finally, there is ground for believing that a number of governments are attempting to change the international image of their countries' investment climates.

V. Official Policies to Promote Investment in Nonfuel Minerals in Developing Countries

This chapter summarizes some of the recent activities and proposals for action by national governments in the developed countries and by international agencies to promote investment in the nonfuel minerals industries of the developing countries.

POLICIES OF DEVELOPED-COUNTRY GOVERNMENTS

It should be said at the outset that the positions of the major developed countries often differ regarding the desirability of promoting mineral production in the developing countries. On the one hand, some public and business opinion in those developed countries that are important producers of nonfuel minerals tends to fear foreign competition. This is particularly true of the United States and Canada.[1] U.S. and Canadian mining companies are also generally opposed to financial and technical assistance by international agencies to public-sector mining enterprises in developing countries.[2] In the United States at least, some labor unions and political circles oppose foreign direct investment in general because of its possible adverse effects on U.S. employment. There is also opposition (e.g., the Mining and Minerals Policy Act of 1970) to increased dependence on foreign supplies of minerals which the United States is capable of producing domestically.

On the other hand, Japan and the countries of Western Europe, which depend heavily upon foreign supplies of nonfuel minerals, have strongly favored foreign mineral development. In fact, their

---

1 See, for example, petitions submitted recently to the U.S. International Trade Commission by 12 U.S. copper producing companies and 7 U.S. zinc producing companies seeking temporary restrictions on U.S. imports of copper and zinc. The petition for a tariff-rate quota on zinc was made on December 20, 1977 and that for an import quota on copper was made on February 23, 1978. The ITC rejected the petition for a tariff on zinc, but recommended an import quota for copper. The latter was rejected by President Carter in October 1978.

2 See, for example, the statement of policy by the American Mining Congress, September 1978: "The American Mining Congress believes that the World Bank, the IMF, and other developmental or financing agencies should not make loans in support of mineral development projects that do not meet usual market criteria of economic viability, or that would aggravate present or projected market surpluses or jeopardize the profitability of existing privately financed mineral projects. Especially in the case of loans for mineral development by state enterprises, the AMC urges that the above criteria be applied rather than political or developmental considerations. The AMC urges the U.S. Administration to support the above policy and, if necessary, to withhold U.S. funds from any such agency that acts incompatibly with this policy which is fundamental to the viability of the nation's mineral and metal industries."

governments have actively supported investment in these industries by subsidizing private investment in the developing countries and in some cases by assisting government mining enterprises there. These countries have also tended to favor international agency assistance for the development of the minerals industries in the developing countries.

Individual OECD governments have devised a number of programs to promote mineral investments in developing countries. These include insurance of equity and loan financing of foreign direct investments against expropriation, war-revolution-insurrection, and currency inconvertibility; the negotiation of investment agreements with the governments of developing countries; government loans and technical assistance to mining enterprises operated by host governments; and long-term government contracts for the purchase of minerals from new mining ventures in the developing countries. This study is not the place for a comprehensive discussion of these programs, but a few comments may be appropriate on the adequacy of certain of the programs and their effectiveness in promoting investment in new mineral producing capacity.

Bilateral Investment Insurance

The United States, Japan and several West European countries have programs for insuring foreign investments by their nationals against various types of political or other noncommercial risks. This discussion is limited to one of these programs, the U.S. Overseas Private Investment Insurance Corporation (OPIC). Its operations have perhaps received the greatest attention because of the large volume of investment that it has insured, particularly in the minerals field. Although in recent years the number of new policies covering mineral investments has been relatively small, OPIC hopes to expand its coverage of new mineral investments under the authority provided in the Overseas Private Investment Corporation Amendment Act of 1978.

Because of widespread expropriation of foreign mineral investments in recent years, it is widely believed that expropriation insurance removes an important deterrent to foreign investment in minerals, although there is also ground for believing that it does not. In any case, many mining firms regard the violation of contract terms short of outright expropriation as constituting an even greater risk than expropriation. Until recently, so-called creeping expropriation, arising from host-government actions that restrict the profitability or limit the control of an investment, was not specifically covered by OPIC. Under a new program adopted in 1978, however, insurance coverage may be designed for host-government breaches of contract that prevent or impair profitability of operations, such as new taxes or government control over marketing. OPIC will cover losses arising from a breach of contract for a two-year period, following which the investor would have the option of closing down the operation and recovering the insured book value as of the date of the breach, less the amount of any compensation paid by OPIC during the two-year period.[3]

Some critics of government insurance of foreign investment have regarded such programs as an improper intrusion by the foreign investor's government into conflicts between the foreign investor and the host-country government. It is argued that political and commercial relations between countries will be influenced by investment disputes in a manner harmful to the mutual interests of the countries involved. Other critics argue that government-sponsored investment insurance constitutes a subsidy to transnational corporations that cannot be justified in terms of the national welfare.

In answer to the first criticism, it is argued that the home government of a foreign investor inevitably becomes involved in an expropriation or breach of contract. (In a sense, the U.S. government is financially involved in a loss by a U.S. corporation as a consequence of expropriation since the loss is written off against the corporation's U.S. tax obligations.) OPIC has argued that the existence of expropriation insurance fosters the settlement of disputes between a foreign investor and a host government. The host government is more willing to reach an agreement with the investor on compensation, since OPIC (that is, the U.S. government) assumes the claim of the foreign investor against the host government if OPIC compensates the investor.

The argument that government investment insurance is a form of subsidy is more complicated. OPIC correctly asserts that its insurance premiums collected have exceeded its claim settlements and administrative expenses over the life of the program, and that OPIC has built up a large reserve.[4] However, its potential liabilities arising

---

3 See *Investment Insurance Handbook: New Programs for Minerals and Energy* (Washington, D.C.: OPIC, November 1977). See also an address by Rutherford M. Poats (Acting President of OPIC) before the American Mining Congress, Phoenix, Arizona, 1978, for a description of new OPIC programs for minerals industries.

4 OPIC's basic premium rates for insured equity investments in mining projects are 0.9 percent per year for expropriation, 0.6 percent for war, revolution and insurrection, and 0.3 percent for inconvertibility. However, higher rates may be charged according to the risk for specific projects.

from insurance contracts are far in excess of its reserves accumulated from net earnings—as is normally the case for any private insurance company. Although life and property insurance contracts are based on actuarial principles, there is no generally accepted actuarial basis for political risk investment insurance. OPIC does reinsure a portion of its portfolio with private insurers, such as Lloyds of London, but OPIC's entire risk exposure could not be transferred to the private market except at substantially increased premiums—probably too high to attract foreign investors. Therefore, investment insurance may be regarded as a subsidy, just as the absorption of political risk on loans by the Export-Import Bank or the World Bank may be regarded as a subsidy.[5]

As of September 30, 1978, OPIC's management estimates its maximum potential exposure on claims arising out of political risk investment insurance to be $3.2 billion (excluding obligations under guarantees issued in settlement of claims). On that same date, OPIC's insurance and guaranty reserves totaled $383 million.[6]

OPIC's insurance program for mineral projects is limited in several ways, including a limitation of $150 million on combined equity and loan guarantees for any one project. This is much too small to cover large mine projects with total required investment running up to a billion dollars or more. In addition, OPIC insurance is not available for investment in certain countries and for certain commodities.

Whether OPIC's political risk investment insurance is regarded as a subsidy, any government service not available in the private market on comparable terms should be justified in terms of its contribution to the national welfare, including a country's interest in promoting the welfare of the world economy. Any activity that removes artificial barriers to the flow of capital for the development of the world's resources might be justified as contributing to the national welfare.

Multinational insurance, which is discussed below, might offer some advantages over bilateral insurance programs. But no investment insurance program can deal with all of the deterrents to foreign investment in the minerals industries. Foreign investors are often

---

5 OPIC officials deny that their operations involve a subsidy. This is a matter of definition. Economists generally define a subsidy as the provision by the government of any service at a price less than that available in the private market.

6 See *Overseas Private Investment Corporation Annual Report for 1978* (Washington, D.C.: OPIC, 1978).

faced with a number of minor breaches of contract which impair profitability, none of which may be sufficient to cause a withdrawal from the project or provide a basis for damage covered by insurance. More importantly, foreign investors are not satisfied simply with an assurance that they will recover their investment; they want reasonable assurance that they will be able to retain a successful investment and to realize returns that will compensate them for other investments that do not prove to be commercially viable. The loss of an investment after a firm has devoted its resources in the form of skilled and experienced personnel to the project over a number of years means that the firm has given up other opportunities for investment which would have contributed to growth and profits.

Bilateral Investment Treaties

Several European countries, including France, Germany, the Netherlands, and the United Kingdom, have negotiated specialized bilateral investment treaties with a number of developing countries covering such important issues as nondiscriminatory or national treatment, expropriation and arbitration of investment disputes.[7] The United States has also negotiated with developing countries a number of general Treaties of Commerce, Friendship and Navigation, but this program, which was rather active during the earlier postwar period, has tended to lag, in part because of the failure of the Congress to ratify treaties entered into by the Administration. The last such U.S. treaty with a developing country to be ratified was that with Thailand in June 1968. One is now under negotiation with Singapore.

Investment treaties tend to be drawn up in rather broad terms, and while they do provide an important framework for negotiations relating to investment disputes, they offer little in the way of legal protection to foreign investors. An important component of any investment treaty should be an agreement to submit disputes arising out of foreign investments to international arbitration—either through the World Bank's International Center for the Settlement of Investment Disputes, the International Chamber of Commerce Court of Arbitration or some other international tribunal. However, host countries usually reserve the right to determine which investments shall be subject to arbitration or which provisions of investment agreements shall be subject to international arbitration.

---

7 For a description of these treaties, see *Bilateral Treaties for International Investment* (Paris: International Chamber of Commerce, January 1977).

Bilateral Loans and Technical Assistance to National Mining Enterprises

Several European governments have provided financial and technical assistance to government mining enterprises in developing countries, partly as a means of promoting the development of foreign mineral supplies. The government of the Federal Republic of Germany or its agencies have been especially active in this area. Such activities might be criticized as favoring public over private ownership and control over minerals resources, although this would hardly apply to the West German government. Such assistance has not been significant in the U.S. Agency for International Development program in recent years, though the Canada Development Corporation has provided some.

Long-term Government Contracts for the Purchase of Minerals

During the 1950s, the U.S. Defense Materials Procurement Agency (DMPA) negotiated long-term contracts with new foreign mining ventures, partly as a means of ensuring adequate supplies of minerals for U.S. defense use and for the U.S. stockpile, and partly to encourage an expansion of foreign capacity for minerals in which the United States was not self-sufficient. The U.S. government also made loans to facilitate the development of new mineral capacity both within the United States and abroad. More recently, there is no evidence of the government's using procurement for the U.S. stockpile to promote the development of new capacity either at home or abroad. The governments of Japan and of certain West European countries, however, have directly or indirectly employed long-term purchase contracts to promote the creation of foreign capacity for the production of raw materials.

The issues relating to stockpile policy are far too complex to be dealt with in this study. However, there appear to be disadvantages in employing a stockpiling program for the creation of new productive capacity. Investments in new capacity should be made on the basis of long-term market considerations, but government stockpile purchase contracts may well distort perceptions of the long-run growth in demand. Moreover, efforts to encourage the expansion of new overseas capacity may be inconsistent with the national security goals of a stockpiling program.

INTERGOVERNMENTAL ACTIONS TO PROMOTE NONFUEL MINERAL INVESTMENTS

Government officials of a number of industrialized countries and the secretariats of several intergovernmental agencies have emphasized the need to promote an adequate level of investment in the minerals industries of the developing countries through international action.

This issue was dramatized by Secretary Kissinger's proposal for an International Resources Bank (IRB) at the ministerial meeting of UNCTAD-IV in Nairobi, Kenya in May 1976.[8] The IRB proposal was largely rejected by the Third World members of UNCTAD, principally because of its emphasis on promoting foreign investment in their resource industries. Although the Carter Administration does not support the Kissinger proposal for an IRB, the problem of promoting mineral investments in the developing world has been the subject of continued study, especially within the Treasury and State Departments. In any event, the problem of inadequate investment in minerals industries is widely recognized.

For example, the staff of the World Bank has prepared several internal studies on the technical, financial and managerial requirements of expanding mineral producing capacity in the developing countries and on the possible roles of the Bank in facilitating the transfer of resources for these purposes.[9] These studies no doubt lay behind the World Bank's discussion and statement of policy on "Nonfuel Mineral Development" in its recent annual report for 1978.[10] (See insert on following two pages.)

In 1976, the General Assembly of the Organization of American States (OAS) passed a resolution calling for a study of an Inter-American Resources Financial Mechanism. This was to be presented to the Special OAS Assembly on Development Cooperation in 1977. A further indication of the interest in this problem is found in a pro-

---

8 *Expanding Cooperation for Economic Development*, an address by Secretary Henry Kissinger, May 6, 1976, Nairobi, Kenya (Bureau of Public Affairs, U.S. Department of State, Washington, D.C.), reprinted in *Mineral Development in the Eighties: Prospects and Problems*, a report prepared by a Group of Committee Members (British-North American Committee, November 1977), pp. 27–28.

9 Some of these studies are reflected in the World Bank research publication entitled *The Mining Industry in the Developing Countries* by Rex Bosson and Bension Varon (Oxford University Press, 1977).

10 World Bank, *Annual Report, 1978*, pp. 20–21.

WORLD BANK STATEMENT ON MINERAL AND FUEL DEVELOPMENT

A continuing supply of both nonfuel minerals and sources of energy is critical to the health of the world economy. Their exploitation is of particular importance to those developing countries in which they are found. Nonfuel minerals can be a major source of foreign exchange earnings and domestic revenues. They can also serve to stimulate development in other sectors. The importance to developing countries of cutting back imports of petroleum products is obvious; virtually all developing nations have a strong incentive to develop indigenous sources of energy as rapidly as possible. During the past fiscal year, the Bank's lending policies in both these fields were assessed, and were modified to fit better the needs of the developing countries. Substantial changes in the existing lending patterns will have to looked at, however, in connection with the available resources of the Bank.

Nonfuel Minerals. The views of developing countries and private foreign investors on how to exploit mineral resources have tended, in recent years, to diverge rather than come together. Developing countries that possess reserves of minerals have shown increasing dissatisfaction with the "enclave" arrangements that were typical in the past. Their claims for a greater voice in decisions concerning the exploitation of mineral resources and for a more balanced sharing of the economic and social benefits to be derived from mining ventures are coming to be generally recognized. But foreign mining companies and investors have hesitated to commit large funds to mining ventures located in countries that appear politically unstable or where there is a serious risk that the terms of investment agreements may be changed by the host government. Uncertainty on this score is perhaps the main reason why mineral investment, particularly equity investment, during the last two decades has taken place largely in developed countries.

International financial institutions—the World Bank, IDA, and IFC, together with the regional development banks—can help bridge the differences between producing countries and foreign mining concerns by providing an international "presence" in mining ventures. Agreements governing projects in which one of the international financial institutions is an active partner are more likely to be regarded as fair by all parties and, therefore, to endure. In this way, foreign investment—particularly risk capital—should become more readily available for mineral production in the developing countries than it has been in the past.

In practice, the World Bank's role as an "active catalyst" would be:

—To help prepare projects and provide assistance at an early stage. Such preparation is needed, for mining projects, which typically involve a web of interrelated technical, financial, legal, and commercial arrangements, are among the most complex and costly in the world. By its involvement, the Bank could be in a position to help all parties arrive at agreements that are both fair and that have a reasonable chance of being observed or of being renegotiated under agreed rules; and to encourage the parties to agree on suitable arrangements for the settlement of disputes, including, where appropriate, recourse to ICSID;

—To provide assistance to developing countries in determining their resources, in planning a strategy for resource exploitation, and in obtaining technical expertise to design, implement, and operate mining ventures.

In order to carry out these functions, and to maintain an "effective presence," the Bank must be willing, however, to contribute a significant share of the financing of nonfuel mineral projects. This share might average about 15% of total project costs, but it may vary substantially.

The estimate of gross investments required in the mineral sector in developing countries for the period 1976–85 runs as high as $95,000 million. As much as two-thirds of this amount may have to come from foreign private sources. There is a consensus in the industry that unless expenditure on new mining capacity begins to accelerate soon, supplies of some essential minerals will fall short of demand in the 1980's, since it takes 10 years or more to develop a new mine. At present, the prospects for increased investment in this sector are clouded by the depressed state of the market for most minerals. It may be another year or two before the effects of the present oversupply wear off and investors are willing to consider new ventures.

In the near term, the Bank expects to assist in the financing of two to three projects a year, and to double this number as soon as the demand for minerals has revived. For the reasons mentioned earlier, the Bank's loans and credits for mining projects will be associated with much larger volumes of investment, both foreign and domestic.

Reprinted from World Bank, *Annual Report, 1978,* pp. 20–21.

posal for an International Minerals Investment Trust (IMIT) which was presented by the U.N. Secretary General in May 1977 to the UN Committee on Natural Resources.[11]

We may distinguish between activities of governments or international agencies which promote or facilitate investment in the minerals industries of developing countries generally, and those activities, such as investment insurance, that are specifically designed to encourage foreign direct investment in these industries, although the former type of activities may also serve to promote foreign direct investment. For example, increased loan or equity participation by the World Bank in mineral projects may facilitate both national investment and foreign direct investment in the minerals industries. The international activities discussed in the following paragraphs are mainly those of the second category, intended to expand foreign direct investment in minerals.

Public International Financial Assistance

Neither the World Bank nor the regional public development banks, e.g., the Inter-American Development Bank or the Asian Development Bank, have made a significant proportion of their resources available to the mining sectors of their member countries. During the post-World War II period to mid-1977, the World Bank group, with has been the most active in this area, had made some 34 commitments totaling about $850 million to the mining sectors of developing countries. According to new policies with respect to the extractive industries adopted by the World Bank in 1977, that institution plans to increase its lending for developing-country nonfuel mineral projects to a total of $800 million during the period 1977 through 1980.[12] Under its new policy of expanded support in this area, mentioned above, the World Bank hopes that by its "presence" in mining ventures and by its contributing on the average some 15 percent of the total project capital, it will serve as an "active catalyst" to attract from foreign private sources about two-thirds of the greatly enlarged amounts of capital that will be required by the developing countries to meet estimated new mining capacity.

11 See *Minerals: Salient Issues*, Report of the Secretary-General, United Nations Committee on Natural Resources (Geneva, March 1977), Annex II; see also Steven A. Zorn, "The United Nations Panel on International Mining Finance," *Natural Resources Forum* (April 1978), pp. 291–299.

12 *International Finance*, Annual Report of the National Advisory Council on International Monetary and Financial Policies for 1977 (Washington, D.C., April 1978), p. 62.

Although it is unlikely that public international development institutions can be counted on to provide a large share of the finance required to expand nonfuel mineral capacities in the developing countries, even their modest participation in both loan and equity financing could be a powerful instrument for mobilizing private international loan financing and for giving greater confidence to potential private international lenders. Two reasons account for this. First, some national or private financial institutions rely on the project investigation and evaluation undertaken by the international agency, and, second, there is a general belief that governments are less likely to default on loans in which international agencies have participated. This confidence might be further increased by the use of cross-default obligations which provide that a default to one creditor constitutes a default to all creditors. International agency participation in project formulation and equity financing should give private foreign equity investors considerable security against expropriation or other contract violations which would affect the earnings of the enterprise. In such cases, the initial amount of international agency participation would not need to be a larger share of the total project cost than the anticipated 15 percent.

Both loan and equity investments by the International Finance Corporation in the World Bank group have been a factor in attracting private investment in developing countries, but equity investment has rarely been used by IFC in support of foreign investment in mining. A notable exception was an IFC loan of $26 million plus an equity subscription of $4 million to CODEMIN, a Brazilian nickel mining and refining project sponsored by the Hochschild Group, a South American international mining and metal trading organization. The IFC also recently made three small loans of $15 million each in support of mining operations, one of which represented IFC's participation in a $400 million loan package for the Cuajone copper mine in Peru; another represented a share in the loan financing of a $200 million nickel mining and processing plant in Guatemala sponsored by INCO, Ltd.; and the third was made to a Brazilian company, Mineracao Rio do Norte, S.A., jointly involving Brazilian and foreign equity organized to mine and process bauxite.

The degree to which public international development agencies should be involved in the actual negotiation of mine development agreements has been a matter of some concern to international mining-company officers. For example, the head of a large U.S. mining firm has expressed the view that a three-way negotiation would be cumbersome and that, in any differences between the foreign investor and host government, the international agency might well side with the host country. On the other hand, it might be argued that the international agency would be able to view objectively the conditions necessary to attract foreign direct investment, and that its participation in the negotiations would provide a measure of support to host-government officials who accepted politically unpopular provisions of a mining agreement.

Another objection to public international-agency participation in the financing of mining projects involving foreign direct investment has been mentioned by officers of two international mining firms. An international agency is alleged to have established environmental standards that went beyond those required by the host government or believed warranted by the foreign investor, and these requirements added significantly to the cost of the investment. This raises the question of the propriety of an international agency's establishing its own social criteria in matters concerning the environment, worker safety and other areas not directly connected with the economic feasibility of a project.

International Investment Codes

Several international organizations have been negotiating codes of conduct for transnational corporations and for the treatment of multinational enterprises by the governments of host countries. In June 1976, 23 members of the OECD adopted a "Declaration on International Investment and Multinational Enterprises" which included voluntary guidelines for multinational enterprises (MNEs), a commitment by the OECD governments to accord "national" or nondiscriminatory treatment to MNEs vis-à-vis domestic firms, and a commitment by host countries to treat MNEs in accordance with international law and with contracts to which the governments have subscribed.

Negotiations on an investment code for MNEs involving both developed and developing countries have been in process in the U.N. Commission on Transnational Corporations, the International Labor Organization, the Organization of American States, and other international and regional organizations. Progress in reaching agreement on a code of conduct has been exceedingly difficult because of the wide differences between the developed countries, on the one hand, and most of the developing countries, on the other. The developing countries want a set of legally binding principles which the home governments of the foreign investor as well as the host-country governments would be duty-bound to enforce, while the developed countries insist that any code be voluntary in nature.

Developing countries emphasize the sovereignty of host governments with respect to MNEs while the developed countries have argued for a commitment by the host countries not to discriminate against MNEs as opposed to domestic firms, and to maintain contract provisions.[13]

The outlook for an international investment code that would provide any significant encouragement to foreign investors in developing countries is not promising. In fact, the kind of code advocated by the majority of the Third World countries appears likely to place additional obligations on foreign investors without according them protection for their investments.

Although bilateral investment treaties have considerable advantages since they may provide specific commitments between governments enforceable under international law, multinational agreements involving both developed and developing countries would also offer certain advantages. Even though such agreements might be voluntary in nature and involve commitments only to certain principles, the violation of these commitments by one multinational firm or host government would be regarded as a default on an obligation to all parties to the agreement. However, since it is improbable to achieve an agreement involving a large number of developing countries that would be effective in stimulating the flow of foreign investment, it might be possible for a group encompassing those developed and developing countries that were particularly interested in reducing barriers to the flow of international investment to adopt their own multinational investment code. That such a code might encourage MNEs to invest in those countries that were signatories to the code, as against countries that were not, might serve to encourage wider formal adherence by still other countries.

Multinational Investment Insurance

Multinational investment insurance has been discussed in the World Bank and the Inter-American Development Bank, as well as in the United Nations. It has also been considered by the European Community. Such insurance would have the advantage of spreading the risk among a number of countries and of subjecting governments guilty of expropriation or contract violation without compensation (in an amount at least equal to the claims of the insured firms against the multinational agency) to sanctions by all the members of the

---

13 For a summary of negotiations on codes of conduct for international investment, see ibid., pp. 86–89.

multilateral organization. The problems involved in reaching an agreement on a multilateral insurance organization are similar to those encountered in agreeing on an international investment code. Many developing countries hold to the principle that a sovereign nation has an absolute right to take any action against foreign owned property that it believes to be in its national interest and to determine what compensation, if any, is due under its national laws and judicial procedures. It would appear that any such multinational insurance should be accompanied by an agreement to subject all insured investments to multinational arbitration of disputes between the foreign investor and host government. However, no Latin American country is a member of the World Bank's International Center for the Settlement of Investment Disputes, nor are the governments of several other important nations, including Algeria, India, Iran, the Philippines, and Thailand. It may be noted that a number of new investment agreements in the nonfuel minerals industry have provided for international arbitration of disputes employing arrangements other than ICSID (see Chapter IV).

An alternative approach to multilateral investment guarantees would create an organization composed of the national investment insurance agencies of the OECD countries, or of the EC countries. Such an organization might begin by issuing joint insurance on a case-by-case basis, with the insurance covering the investments made by the nationals of any of the countries involved in the program. This would require a change in the U.S. legislation governing OPIC which covers only U.S. investors and which prohibits OPIC insurance to certain countries and for investment in the production of certain commodities.[14] A multilateral insurance program would be faced with the difficult problem of establishing international standards of compensation in cases of expropriation and breach of contract. Rather than trying to draw up detailed standard contracts before issuing joint insurance, it might begin by experimenting with various types of contracts on a case-by-case basis in order to gain experience. This would be particularly important in insuring against losses arising from breach of contract short of full expropriation.

Consideration of the problem of harmonizing the operations of existing national insurance schemes—including domestic objections

---

14 For example, OPIC is not permitted to insure investments in copper mining projects scheduled to begin operations before January 1, 1981 or after that date if the project would cause injury to the U.S. primary copper industry. Also, OPIC cannot issue insurance on investments in countries that are identified as violating human rights, the selection of such countries having been largely political. See *Overseas Private Investment Corporation Amendment Act of 1978*, Section 7, 95th Cong., 2nd sess., 1978.

to allowing a national insurance agency to insure an investment whose entire output was scheduled for a third country—has led many students of the problem to conclude that only a regional or world multinational investment insurance scheme would be practical. It also seems likely that a multinational scheme limited to those countries that are the major capital exporters would be easier to form than a worldwide scheme including the developing countries.

An International Minerals Investment Trust

In May 1977, a proposal for an International Minerals Investment Trust (IMIT) was presented to the United Nations Committee on Natural Resources.[15] According to this proposal, the assets of the IMIT would consist of both debt and equity issues of companies established under the laws of developing countries and set up specifically to exploit ore bodies in those countries. Seed capital would be provided by the governments of the developed countries and possibly by the OPEC nations, while subsequent financing would be provided by sales of shares of the IMIT to both public and private agencies. Unlike the International Resources Bank which was rather heavily oriented to the financing and guaranteeing of private international mining investments, IMIT would provide a mechanism for financing national mining enterprises in developing countries. The proposal suggests that international mining firms would play a role in the provision of technology and management and possibly as providers of minority equity shares in new projects. However, this does not appear to be a proposal that is likely to prove very interesting to the international mining community or to deal with obstacles to foreign direct investment discussed in earlier chapters.

CONCLUSIONS

Of the various approaches to facilitating the flow of foreign direct investment in the minerals industries of developing countries described in this chapter, participation in both debt and equity financing by existing public international financial agencies such as the World Bank group and the Inter-American Development Bank appears to offer the greatest promise.[16] The amount of the actual

---

15 *Minerals: Salient Issues*, Annex II.

16 Currently, the IDB cannot make equity investments, but an equity affiliate of the IDB has been proposed.

financial participation need not be large since the principal objective would be to provide foreign investors with the protection afforded by having an international public financial agency as a party to the creditor agreement in the case of loans, and to the mine development agreement in the case of equity participation. In negotiating such three-way agreements, the international financial agency's role would appear sounder if it would avoid trying to dictate the terms of mine development agreements or developing rigid standards of its own; and would confine its role largely to technical advice and conciliation.

Appendices to Chapter I

APPENDIX A
Projections of World Demand for Major Nonfuel Minerals

A bewildering number of projections of world demand[1] for major nonfuel minerals have been made by the research staffs of various agencies, including the U.S. Bureau of Mines, the World Bank, the United Nations Center for Natural Resources, Energy and Transport, the Commodities Research Unit, Inc., Charles River Associates, Inc., and the research departments of mining companies, among others. Many of these projections are not comparable because they employ different concepts of consumption (e.g., refined versus total copper consumption) or because of differences in geographical coverage, or because of the use of different base years. However, even if these difficulties were removed by adjustments to comparable bases, there would remain among these expert projections a fairly wide range of annual rates of growth of consumption between those known from recent historical rates and those projected between now and the end of this century.

Table A-1 compares historical rates of growth in world demand for five nonfuel minerals with those projected growth rates to the years between 1990 or 2000 as estimated by the U.S. Bureau of Mines, the World Bank and by Professor Wilfred Malenbaum of the University of Pennsylvania. This material is summarized in Chapter I, Table I-1.

Taking into account the differences in base periods employed, the U.S. Bureau of Mines and World Bank projections are not too dissimilar for iron ore and tin, but they differ considerably for aluminum, copper and nickel. The estimated rates of growth in world demand for refined copper are higher in the case of the U.S. Bureau of Mines, and the projected rates of growth in the demand for aluminum and nickel are significantly higher in the case of the World Bank staff. Except for tin, Malenbaum's projections are substantially lower than those of both the U.S. Bureau of Mines and the World Bank. Malenbaum's 1978 study[2] is a revised version of his study prepared for the U.S. National Commission on Materials Policy in 1973.[3] His 1978 projections show substantially lower average annual growth rates of consumption to the year 2000 for nearly all the nonfuel minerals than those he had offered in his 1973 study.

With the exception of tin, all of the projections shown in Table A-1 indicate lower average annual rates of growth in demand than their actual growth rates over the past two decades.[4] An important contributing factor is lower projected growth in

---

1 Unless otherwise indicated, the term "world" excludes the Soviet bloc countries and Mainland China.

2 Wilfred Malenbaum, *World Demand for Raw Materials in 1985 and 2000* (New York: McGraw Hill, 1978).

3 Wilfred Malenbaum, *Materials Requirements in the U.S. and Abroad in the Year 2000* (Philadelphia: University of Pennsylvania, March 1973).

4 With the exception of the Bureau of Mines projections for copper, the demand projections given in Table A-1 for aluminum, refined copper, nickel, and tin refer to consumption of the metal rather than to the demand for primary metal. If an increasing proportion of the total demand were supplied from recycling, the demand for primary materials would be reduced. For example, the increased proportion of aluminum used for canning and other packaging has shortened the life cycle of aluminum products, and therefore increased the proportion of recycled aluminum.

TABLE A-1: HISTORICAL AND PROJECTED RATES OF GROWTH IN WORLD DEMAND FOR SELECTED NONFUEL MINERALS
(Annual averages over the periods indicated)

	Historical[1]	U.S. Bureau of Mines[2]	World Bank Staff[1]	Malenbaum[3]
Aluminum	7.3% (1960–76)	5.2% (1975–2000)	6.7% (1974/76–90)	3.0% (1975–2000)
Refined copper	3.9 (1955–77)	4.0 (1975–2000)[a]	3.4 (1977–90)	2.1 (1975–2000)
Iron ore	3.6 (1960–76)	2.8 (1973–2000)	3.2 (1976–90)	2.1 (1975–2000)
Nickel	6.5 (1950–74)	3.5 (1975–2000)	5.1 (1976–90)	2.1 (1975–2000)
Tin	1.0 (1955–76)	1.5 (1973–2000)	1.3 (1974/76–90)	1.7 (1975–2000)

a World demand for primary copper (excluding scrap). This is a composite of a 4.5 percent projection from 1975 to 1985 and 3.7 percent from 1985 to 2000.

1 Data from internal World Bank staff study; data exclude centrally planned economies.

2 Copper forecast: H.S. Schroeder, *Copper* (Bureau of Mines, June 1977), p. 14; Nickel forecast: J.D. Corrick, *Nickel–1977* (Bureau of Mines, July 1977), p. 15; Aluminum, iron ore and tin forecasts: *Mineral Facts and Problems* (Bureau of Mines, 1976), pp. 60, 543 and 1,137.

3 Wilfred Malenbaum, *World Demand for Raw Materials in 1985 and 2000* (New York: McGraw Hill, 1978), Table 5a.

the gross domestic product (GDP) of the developed countries. In addition, Malenbaum believes that as per capita GDP grows, the intensity of use of minerals (i.e., the amount of material consumed per unit of GDP) first increases in a given economy, then levels off and finally declines. The forces responsible for the declining intensity of use include (a) shifts in the types of final goods and services demanded directly by consumers and investors; (b) technological developments that alter the efficiency with which raw materials are discovered, extracted, processed, and utilized in the production of final goods; and (c) substitution among raw material inputs in response to relative price movements and relative rates of technological development.

APPENDIX B
World Mineral Resources and Reserves

Supply Availability

There are several dimensions of supply in relation to world demand for a mineral resource. First, on a purely physical level, we can estimate the total volume of the mineral in the earth's crust plus the amount of the mineral contained in all of the goods thus far produced by people. This constitutes the absolute maximum availability of the mineral. However, in order to become available for commercial use, a mineral in the earth's crust must be sufficiently concentrated to equal or exceed the minimum cutoff grade that is economically recoverable. Thus, for example, the average crustal concentration of copper is estimated to be about 63 parts per million (0.0063 percent). However, in 1975, the lowest "cutoff" concentration of copper economically recoverable in an ore body without a by-product was about 0.35 percent, or 56 *times* the crustal concentration of copper. For mercury, the corresponding ratio is 11,200 times the average crustal composition; but for aluminum it is only about 2 times (see Table B-1). These ratios tell nothing about the distribution within the earth's crust of higher-than-average concentrations of minerals. To put it another way, if there were no concentrations of copper in the earth's crust substantially higher than the average concentration, copper could never have become an economic metal.

A more limited but more useful concept of the world's resources of a metal—for instance, copper—includes all the naturally occurring copper-bearing materials that exist in the earth's crust in such form that economic extraction is currently or potentially feasible. Such minerals resources are usually divided into "identified" resources and "undiscovered" resources. That portion of identified resources from which the metal can be economically extracted at the time of determination is called a mineral reserve, or sometimes an ore deposit.[1]

The volume of world reserves of a metal that can currently be extracted economically is a function of a number of factors: the price of the metal, existing technology, the cost of exploring, extracting the ore from the ground, and extracting the metal from the ore and converting it into a commercially salable product. There are several aspects of exploration, ranging from the mere identification of a deposit to the determination with reasonable accuracy of the economic feasibility of production that will cover the operating and capital costs of mining and processing the metal. Hence, the volume of reserves of a metal rises or falls with changes in its price, with new technological developments, with changes in the costs of inputs, including energy, and with the amount and success of exploration activities.

---

1 Identified copper reserves are divided into three classes: measured, indicated and inferred, depending on the amount of information existing about the material. An identified copper reserve is a body of copper-bearing material whose location, quality and quantity are known from geologic evidence supported with engineering measurements, while an undiscovered copper resource is a body of copper-bearing materials surmised to exist on the basis of broad geologic knowledge and theory. Undiscovered copper resources are divided into "hypothetical" resources in known copper-bearing districts, and "speculative" resources in undiscovered districts. Hypothetical and speculative resources may be added to the volume of world reserves when sufficient information about them is obtained through exploration.

The above definitions are taken from *Geological Survey Bulletin*, 1450-A, published jointly by the U.S. Bureau of Mines and the U.S. Geological Survey (Washington, D.C., 1976). See also *Mineral Development in the Eighties: Prospects and Problems*, a report prepared by a Group of Committee Members (British-North American Committee, November 1977), pp. 30–31.

TABLE B-1: RATIO OF CUTOFF GRADE TO CRUSTAL ABUNDANCE FOR SELECTED ELEMENTS (LOWEST CONCENTRATION ECONOMICALLY RECOVERABLE IN 1975)

Element	Ratio
Mercury	11,200-to-1
Tungsten	4,000
Lead	3,300
Chromium	2,100
Tin	2,000
Silver	1,330
Gold	1,000
Molybdenum	770
Zinc	370
Uranium	350
Carbon	310
Lithium	240
Manganese	190
Nickel	100
Cobalt	80
Phosphorus	70
Copper	56
Titanium	16
Iron	3
Aluminum	2

Source: Earl Cook, "Limits to Exploitation of Nonrenewable Resources," *Science* (Vol. 191), February 20, 1976, p. 678.

Over the past quarter century, world reserves of the principal nonfuel mineral commodities have actually been rising. Indeed, additions to reserves have exceeded by several times the cumulative production during the same period (see Table B-2). The principal factor behind increasing reserves has been exploration, but in some cases technological advances have played an important role in reducing cutoff grades. Exploration and technological advance are dependent on economic factors such as cost-price relationships and the growth in demand. Exploration is a high-risk form of investment and is sensitive to long-run profit expectations. Likewise, the development and utilization of new technology are functions of investment decisions in response to profit opportunities. Technological advance has played a major role in keeping real costs and prices of nonfuel minerals relatively stable in the face of declining grades of ore mined.

The adequacy of reserves for meeting future demand is sometimes assessed by calculating the ratio of existing reserves to forecasts of cumulative demand for a given period in the future. Estimates of these ratios for the years 1974 to 2000 are given for 13 metallic minerals in Table B-3. Only in the case of mercury is the ratio of identified *reserves* to projected cumulative demand less than unity, but for identified *resources* (which include deposits known to exist but subeconomic to recover at base-year prices), the ratio is above unity for all the minerals listed. Table B-4 indicates the life expectancy of reserves, estimated as of 1974, at four assumed growth rates of demand, compared to the actual average annual growth rates for the period 1947–74. This table suggests that reserves of mercury, silver and zinc might not be sufficient to support production of these minerals at historical growth rates.

TABLE B-2: CUMULATIVE PRODUCTION AND ADDITIONS TO WORLD RESERVES, SELECTED MINERAL COMMODITIES, 1950-74
(Metric tons)

Mineral	1950 Reserves	1950-74 Cumulative Production	1950-74 Addition to Reserves[1]	Additions to Reserves as a % of 1950 Reserves[2]
Bauxite	1.4 billion	.85 billion	15 billion	1,103%
Chromium	100 million	96 million	1.7 billion	1,696
Cobalt	79,000	440,000	2 million	258
Copper	100 million	110 million	400 million	403
Iron	19 billion	7.3 billion	76 billion	401
Lead	40 million	63 million	170 million	433
Manganese	500 million	160 million	1.6 billion	313
Mercury[3]	130,000	190,000	2.5 million	188
Molybdenum[4]	4 million	1.1 million	2.0 million	51
Nickel[4]	14 million	9.4 million	39 million	281
Platinum group	780	1,700	20,000	2,564
Silver	160,000	200,000	230,000	141
Tin	6.0 million	4.6 million	8.6 million	144
Tungsten	2.4 million	760,000	– 43,000	– 1.8
Zinc	70 million	97 million	150 million	210

1 Derived by adding 1950-74 cumulative production to 1974 reserves and subtracting 1950 reserves.
2 Calculated before reserve and cumulative production data were rounded.
3 Reserve figure for 1950 is incomplete for Communist countries.
4 Reserve figure for 1950 excludes Communist countries.
Source: John E. Tilton, *The Future of Nonfuel Minerals* (Washington, D.C.: The Brookings Institution, 1977), Table 2-2, p. 10.

TABLE B-3: INDICATORS OF PHYSICAL ADEQUACY OF 13 METALLIC MINERALS

		Identified Reserves			Identified Resources		
		Ratio to Cumulative Demand 1974–2000			Ratio to Cumulative Demand 1974–2000		
	Percent Increase 1973–75	U.S.	Other	World	U.S.	Other	World
Iron and alloying metals							
Chromium	4%	—	7.2	5.7	0.1	10.0	10.0
Cobalt	− 1	—	3.0	2.1	2.1	4.4	3.7
Iron ore	4	1.3	5.1	4.5	6.0	10.0	9.8
Manganese	308	—	5.6	4.9	1.6	9.7	8.8
Molybdenum	40	2.0	1.2	1.4	10.0	4.7	7.0
Nickel	29	—	2.7	2.1	2.1	4.9	4.2
Tungsten	42	0.3	1.5	1.2	1.2	4.2	3.5
Nonferrous metals							
Bauxite	10	—	5.9	4.0	0.2	9.7	6.6*
Copper	22	1.2	1.3	1.3	5.3	6.0	5.8*
Gold	35	0.5	1.4	1.3	1.1	2.0	1.8
Mercury	− 1	0.3	0.8	0.7	0.7	3.1	2.7
Platinum group	− 10	—	4.2	3.1	4.4	10.0	8.7*
Tin	144	—	1.7	1.3	0.1	3.3	2.7

Note: Definitions used: "identified reserves" are known deposits economically feasible to recover at 1974 prices; "identified resources" are these reserves plus deposits known to exist, but subeconomic at 1974 prices. "Cumulative demand" covers that for primary metal only; no scrap.

* Figures for resources of bauxite, copper and platinum include hypothetical and speculative resources as well as those presently identified.

Source: Sperry Lea, "Uncertainties in Future Metals Supplies," *New International Realities* (Washington, D.C.: National Planning Association, October 1976), Table 1, p. 3.

TABLE B-4: LIFE EXPECTANCIES OF WORLD RESERVES AS OF 1974, SELECTED MINERAL COMMODITIES

Mineral[1]	Life Expectancy in Years, at Four Growth Rates[2]				Average Annual Production Growth, 1947-74 (percent)
	0%	2%	5%	10%	
Bauxite (ore)[3]	226 yrs.	86 yrs.	51 yrs.	33 yrs.	9.8%
Chromium (ore)	263	93	54	35	5.3
Cobalt (Co)	97	54	36	25	5.8
Copper (Cu)	56	38	27	20	4.8
Iron (Fe)	167	74	46	30	7.0
Lead (Pb)	42	31	23	17	3.8
Manganese (Mn)[4]	190	79	48	31	6.5
Mercury (Hg)	19	17	14	11	2.0
Molybdenum (Mo)	70	44	31	22	7.3
Nickel (Ni)	67	43	30	22	6.9
Platinum group (metal)	117	61	39	27	9.7
Silver (Ag)	20	17	14	12	2.2
Tin (Sn)	42	31	23	17	2.7
Tungsten (W)	42	31	23	17	3.8
Zinc (Zn)	21	18	15	12	4.7

1 The notation in parentheses following the name of a mineral commodity indicates what the reserve and production figures actually measure, e.g., contained copper (Cu).

2 Life expectancy figures were calculated before reserve and average annual-production data were rounded.

3 Figures include only those ores that contain 52 percent or more aluminum.

4 Concentrate is assumed to contain 46 percent manganese for the production figure and 35 percent for the reserve figure.

Source: John E. Tilton, *The Future of Nonfuel Minerals* (Washington, D.C.: The Brookings Institution, 1977), Table 2-1, pp. 6-7.

Such calculations are often improperly used as a basis for forecasting absolute supply shortages or sharply rising prices of the metals in question. The volume of reserves is not static, and any tendency for cumulative production to grow faster than additions to reserves will normally be reflected in price rises. Increased prices spur exploration activities, render a larger proportion of identified resources profitable for exploitation, and a larger proportion of identified resources may also reduce the rate of growth in demand by encouraging the substitution of cheaper materials. These factors have led most mineral economists to reject the concept of absolute depletion of resources in the foreseeable future. Of course, both the likelihood of finding additional reserves and the ease of substituting other materials for their use differ substantially from mineral to mineral.

Estimates of identified but subeconomic resources and of hypothetical and speculative resources usually run several times the volume of reserves. For example, world reserves of copper were estimated by the U.S. Bureau of Mines to be 460 million tonnes (metric tons) in 1976, but over 700 million mt have been identified as subeconomic deposits, about one-half in the form of deep-sea nodules. Hypothetical resources located near known deposits account for another 440 million mt of copper; and yet another nearly 300 million mt are listed as speculative. Much larger reserves of most minerals could readily be proven if additional drilling were undertaken on deposits that are already under exploitation; but, since drilling is expensive, mining companies do not ordinarily prove up reserves many years in advance of the time they expect to exploit them. Hence, estimates of reserves tend to underestimate greatly the total volume of resources that could be produced economically at current prices.

Distribution of Reserves and Production

Tables B-5 and B-6 show the share of world production and reserves of 14 nonfuel minerals by major geographical and political groupings, including, in this case, the Communist countries.

Reserves and other minerals resources—identified, hypothetical or speculative—are a function of the intensity of geological survey and exploration. Since these activities have been undertaken much more intensively in the developed countries than in the developing countries, the geological distribution of reserves given in Table B-5 probably underestimates minerals resources in the developing countries.

Of the minerals listed in Tables B-5 and B-6, the OECD countries consume about two-thirds of the world's output; these countries produce more than 75 percent of the world's output of molybdenum and over 40 percent of the world's output of lead, nickel and zinc. The OECD countries have over 50 percent of the world reserves of mercury, lead, zinc, and molybdenum. South Africa holds over one-half the world's reserves of the platinum-group metals.

The OECD countries are most heavily dependent on supplies from the rest of the world for bauxite, chromium, cobalt, manganese, tungsten, the platinum group, and tin; they are less dependent on the rest of the world for supplies of iron, lead, nickel, molybdenum, copper, and zinc. A substantial proportion of the net OECD imports of chromium, manganese and the platinum-group metals comes from South Africa.

Eight of these nonfuel minerals—bauxite, alumina, copper, iron ore, nickel, tin, lead, and manganese and zinc—account for the vast bulk of imports by the OECD countries. In 1970, the value of the net imports of these commodities by the OECD countries totaled about $5.8 billion, of which $5.0 billion, or 86 percent, were comprised of bauxite and alumina, copper, iron ore, nickel, and tin. The developing countries had net exports of nearly $4 billion of these five commodities. Since 1970, prices and quantities imported of these commodities have changed considerably, but their relative importance in the total net imports of the OECD countries probably has not changed significantly. Special emphasis is given in this study to the supply of these commodities, especially in Appendix C which follows.

TABLE B-5: SHARES OF WORLD MINE PRODUCTION OF 14 NONFUEL MINERALS, 1976
(Percent)

Commodity	OECD	South Africa	Yugoslavia	Developing Countries	Communist Countries	Other Non-Communist Countries*
Bauxite	39.3%	—	2.5%	44.1%	14.1%	—
Chromium	7.9	27.0%	—	11.2	34.8	19.1%
Cobalt	15.1	—	—	71.5	9.6	3.8
Copper	33.7	2.5	1.5	39.0	22.8	0.5
Iron ore	35.1	—	—	17.6	37.0	10.3
Manganese	6.6	23.3	—	22.1	41.7	6.3
Mercury	37.5	—	4.9	17.9	39.5	0.2
Lead	44.8	—	3.4	19.5	30.3	0.9
Molybdenum	75.7	—	—	11.3	12.7	0.3
Nickel	47.1	2.9	—	26.1	23.0	0.8
Patinum group	8.0	45.7	—	0.4	45.5	0.4
Tin	7.0	1.1	—	73.2	18.7	—
Tungsten	17.7	—	—	19.9	49.0	13.4
Zinc	46.9	1.3	1.8	20.3	27.4	2.3

* Data includes production of non-Communist countries not included in data available for OECD, South Africa, and developing countries categories.

Sources: *Commodity Data Summaries, 1977* (Bureau of Mines, 1977); *Yearbook, 1977*, American Bureau of Metal Statistics; and *Metal Statistics 1966–76* (Frankfurt, Germany: Metallgesellschaft, 1977).

TABLE B-6: SHARES OF WORLD MINE RESERVES OF 14 NONFUEL MINERALS, 1975
(Percent)

Commodity	OECD	South Africa	Yugoslavia	Developing Countries	Communist Countries	Other Non-Communist Countries
Bauxite	31.1%	—	1.0%	61.3%	2.6%	3.0%
Chromium	0.7	74.1%	—	22.2	1.0	2.0
Cobalt	8.1	—	—	42.5	22.1	27.3
Copper	28.9	—	—	37.7	8.9	24.5
Iron ore	34.5	1.2	—	28.1	34.2	2.0
Manganese	8.0	44.7	—	8.9	38.3	—
Mercury	62.6	—	10.1	7.0	20.3	—
Lead	68.1	—	1.8	13.3	16.0	0.8
Molybdenum	63.9	—	—	17.0	19.0	0.1
Nickel	25.7	—	—	59.0	15.3	—
Platinum group	1.4	83.4	—	—	15.1	—
Tin	5.4	—	—	64.7	29.9	—
Tungsten	20.9	—	—	10.1	69.0	—
Zinc	69.1	—	2.4	18.0	10.5	—

Sources: *Mineral Facts and Problems: Bicentennial Edition* (Bureau of Mines, 1975), Bulletin 667; *Mineral Commodity Summaries*, 1978 (Washington, D.C.: U.S. Department of Interior, 1978).

APPENDIX C
Aspects of World Supply and Demand: The Cases of Five Major Nonfuel Minerals

Copper

World mine *production* of copper in non-Communist countries was 6.2 million mt in 1976 and 6.3 million mt in 1977.[1] Estimated world copper mine *capacity* at the end of 1976 was 7.5 million mt.[2] On the basis of known expansion plans and an estimate of their realization, one projection for productive capacity at the end of 1981 is 8.1 million mt,[3] an increase of only about 1.5 percent per year over the 1977 to 1981 period. Assuming the usual 93 percent average utilization rate in 1981, the projected capacity expansion would accommodate an annual rate of growth in the demand for primary copper of about 4 percent, excluding any allowance for inventory liquidation or mine closings. Much of the projected capacity expansion is accounted for by Canada, Iran (Sar Cheshmeh), Mexico (mainly La Caridad), Peru, the Philippines, the United States, and Zaire. Some of these projects have been completed, others are well under way, and some (those in Iran and Zaire) face political uncertainties.

Beyond 1981, there are several large copper mine projects that are in the intensive exploration stage for which financing has not been arranged. These include the Ok Tedi project in Papua New Guinea, Pachon in Argentina, Cerro Colorado in Panama, Andacollo and Quebrada Blanca in Chile, and several large copper deposits in Peru.

Each of these projects could have a capacity of 100,000 mt or more per annum, involving a capital cost of over $500 million per project. Few if any of them are likely to begin production much before 1985. Political risk as well as market outlook will play an important role in determining which of these projects will go forward. Tentative mining agreements with foreign mining companies have been negotiated for the development of some of them, including Ok Tedi, Andacollo, Quebrada Blanca, and Cerro Colorado, but in most cases the feasibility studies have not been completed.

It may be noted that well over one-half of the projected increases in copper mine-producing capacity between 1977 and 1981 are located in the developing countries, as are most of the plans to increase capacity that have been announced for completion beyond 1981. The realization of these new projects in the developing countries is by no means ensured. However, if some of the large planned projects in the developing countries do not materialize, a number of opportunities for new capacity exist, both new mines and additions to existing mines in Canada and the United States, and these could be exploited fairly rapidly if markets warranted and/or if other conditions necessitated.

---

1 *Metal Statistics, 1966–77* (Frankfurt: Metallgesellschaft, 65th edition, 1978), p. 30.

2 RTZ Mine Information System; this figure includes Cuajone in southern Peru, which did not come into commercial production until 1977.

3 Controller's Department, Phelps Dodge Corporation (January 1978). Some other estimates run somewhat higher.

TABLE C-1: REQUIRED WORLD COPPER PRODUCING CAPACITY, 1985 AND 2000
(1,000 metric tons per year)

	Historic	Projected	
	1976	1985	2000
Refined copper consumption (Assumed average annual growth rate, 3.5%)	6,417[a]	8,746	14,650
Secondary refined consumption	986[b]	1,300[b]	1,818[b]
Primary refined production	5,431	7,446	12,832
Required *mine capacity* (Allowing for 93% capacity utilization[c] and 3% efficiency factor)	7,530	8,191	14,115
Required *smelter capacity*[d] (Ratio smelting capacity to mine capacity 1.11)	7,830	9,092	15,668
Required *refining capacity*[d] (Ratio refining to mine capacity -1.17)	8,320	9,583	16,515

a *Metal Statistics, 1966–76* (Frankfurt, Ger.: Metallgesellschaft, 1977).
b Joseph F. Shaw, "Investment in the Copper Industry," *Natural Resources Forum* (January 1978), p. 106. His estimates were based on a study by Commodity Research Unit, Ltd.
c Based on average historical ratio of utilization of world capacity over past 25 years, as computed by AMAX Inc.
d Ratios taken from Shaw, *Ibid.*, p. 107

The three projections for world demand for copper shown in Appendix A, Table A-1, range from 2.1 to 4 percent growth per annum. If we assume a constant annual rate of growth of 3.5 percent for refined copper consumption over both the 1976–85 and 1985–2000 periods, we can project the capacity requirements for world primary copper and copper smelter and refining capacities for the years 1985–2000. These projections based on the constant 3.5 percent growth assumption are set forth in Table C-1.

Based on the capacity estimates developed in Table C-1, Table C-2 shows required *additions* to these capacities between 1977 and 1985 and between 1985 and 2000. As compared with its 1976 level of 7.5 million mt, required total mine capacity would nearly double to 14.1 million mt by 2000. The *additions* of 8.4 million mt required between 1976 and 2000 would exceed the total 1976 mine capacity by 11 percent and would also require *additions* to smelter capacity and refining capacity exceeding 1976 capacities by 24 percent and 23 percent, respectively.

Taking Shaw's estimates of the investment cost per annual tonne of new copper capacity, we then calculate in Table C-3 the investment requirements (in 1977 dollars) for creating the additional copper capacity shown in Table C-2. Other industry estimates of investment cost per annual tonne run somewhat higher than Shaw's and a figure of $8,000 instead of his $6,500 per annual tonne for copper (again in 1977 dollars) might be more realistic. Using the higher figure would add over 20 percent to the annual and aggregate total investment requirements estimated in Table C-3.

Over the 1977–85 period, average investment requirements in 1977 dollars are estimated at over $1.1 billion per year, and for the 1985–2000 period, they almost tre-

TABLE C-2: REQUIRED ADDITIONS TO WORLD COPPER CAPACITY, 1977-85, 1985-2000 AND 1977-2000
(1,000 metric tons)

		1977-85 Period (8 Yrs.)			1985-2000 Period (15 Yrs.)			1977-2000 Period (23 Yrs.)		
	Actual Capacity[a] End-1976	Total Closings 1977-85[b]	Required Capacity[c] End-1985	Required Additions 1977-85	Total Closings 1985-2000[b]	Required Capacity[c] End-2000	Required Additions 1985-2000	Total Closings 1977-2000[b]	Required Capacity[c] End-2000	Required Additions 1977-2000
Mine	7,530	678	8,191	1,339	1,129	14,115	7,053	1,807	14,115	8,392
Smelter	7,830	705	9,092	1,967	1,174	15,668	7,750	1,879	15,668	9,717
Refinery	8,320	749	9,583	2,012	1,248	16,515	8,180	1,997	16,515	10,192

a Mine capacity estimate from RTZ Mine Information System; smelter and refinery capacity estimates from Shaw, "Investment in the Copper Industry," p. 108.

b Assumes closings of 1 percent per year on 1976 capacity base for mines, smelters and refineries. This is approximately the same as the estimate for mine closings used by Shaw in "Investment in the Copper Industry," pp. 101-120. However, the 1 percent per annum estimate for smelter and refinery closings is substantially below Shaw's estimates. Shaw's article was based on a U.N. study in which he participated entitled, *Future Demand and the Development of the Raw Materials Base for the Copper Industry*, Report of the Secretary-General (New York, March 1977), mimeo.

c From Table C-1.

TABLE C-3: INVESTMENT REQUIREMENTS FOR ADDITIONS TO COPPER CAPACITY, 1977-85, 1985-2000 AND 1977-2000
(1977 dollars and 1,000 metric tons)

		1977-85 (8 Yrs.)			1985-2000 (15 Yrs.)			1977-2000 (23 Yrs.)		
			Required Investment			Required Investment			Required Investment	
	Investment[a] Cost per Mt per Year ($ mil)	Required[b] Additional Capacity (1000 mt)	Total ($ mil)	Average per Year ($ mil)	Required[b] Additional Capacity (1,000 mt)	Total ($ mil)	Average per Year ($ mil)	Required[b] Additional Capacity (1000 mt)	Total ($ mil)	Average per Year ($ mil)
Mine	$4,040	1,339	$5,410	$ 601	7,053	$28,494	$1,900	8,352	$33,904	$1,143
Smelter	2,000	1,967	3,934	437	7,750	15,500	1,033	9,717	19,434	810
Refinery	460	2,012	926	103	8,180	3,762	251	10,192	4,688	195
Total	$6,500		$10,270	$1,141		$47,756	$3,184		$58,026	$2,418

a Shaw, "Investment in the Copper Industry," p. 109. Assumes a 3:2 ratio of new mines to expansions of existing mines.
b From Table C-2.

ble to average $3.2 billion annually. Over the 23 years, they aggregate $58.0 billion. To these massive capital requirements must be added expenditures for exploration and for pollution abatement. No allowance is made for a rise in the real capital cost per tonne of output, although such a rise seems likely over the next two decades. If copper consumption were estimated to grow by only 3 percent between 1977 and 2000 instead of the 3.5 percent per year used in Table C-1, these total investment requirements in Table C-3 would be reduced by only about 12 percent.

As for the question of where new capacity will occur, a study by three World Bank economists assumes that over 70 percent of the additions to world copper mine and smelter capacity during the 1975–85 period will be located in the developing countries.[4] However, on the basis of current production and political uncertainties, reserve potential and announced plans, it may be questioned whether the developing countries will receive more than about half of the projected additions to copper producing capacity during the remainder of this century. In the actual allocation of investment expenditures between developed and developing countries, even this share will depend upon the ability of the developing countries to obtain the necessary financing and technology, a subject that is discussed in Chapter I.

Bauxite-Alumina-Aluminum

Although the OECD countries produce nearly 40 percent of the world's output and hold about 22 percent of the world's bauxite reserves, they depend substantially upon the developing countries for imports. Projections of the rate of growth of world consumption of aluminum (Table A-1) range from Malenbaum's 3.0 percent per annum through 2000 to the World Bank's 6.7 percent through 1990. Aside from a major expansion now under way in Australia, the largest planned increases in bauxite capacity are located in the developing countries, including Guinea, Brazil, Venezuela, Sierra Leone, and Jamaica. Much of this is being undertaken by joint ventures involving foreign aluminum companies. A large bauxite-alumina expansion in Guinea is being financed by an Arab consortium. Bauxite producing countries are planning to increase their production of alumina, and a number of developing countries, e.g., Brazil, are planning a major expansion in aluminum smelting capacity.

The capital cost of bauxite capacity has been estimated at $75 per annual tonne (in 1975 dollars), and the additional annual capacity required by the market economies between 1977 and 1990 was estimated in an unpublished report prepared for the United Nations at about 39 million mt of bauxite, based on an assumed annual rate of growth in aluminum consumption of 6 percent. Adding consumption growth at an annual rate of 5 percent over the period 1990–2000 would require an additional capacity of about 44 million mt by 2000, or 83 million mt for the 23-year period. Assuming no increase in the real cost of productive capacity over this period, aggregate required investment expenditures for bauxite alone may be projected at $6.2 billion, or about $258 million annually, 75 percent of which, or $194 million per year, may be spent in the developing countries.

The estimated cost of alumina and aluminum capacities per annual tonne (in 1975 dollars) is $600 and $2,400, respectively. Based on the annual rates of growth used above for aluminum consumption (6 percent between 1977 and 1990, and 5 percent between 1990 and 2000), the aggregate cost of additional required alumina and aluminum capacities over the entire 1977–2000 period may be projected at $22 billion

4 K. Takeuchi, G. Thiebach and J. Hilmy, "Investment Requirements in the Nonfuel Minerals Sector of the Developing Countries," *Natural Resources Forum* (April 1977), p. 266.

for alumina and $69 billion for aluminum, or $0.9 billion and $2.9 billion per year (in 1975 dollars), respectively. However, it is expected that the developing countries would account for no more than about 25 percent of the additional capital expenditures for alumina and aluminum capacity.

As for bauxite, we can assume that, as bauxite capacity expands in line with the growth in demand, the bulk of it will be located in the developing countries.

It can be assumed that some international aluminum companies will take advantage of economies in producing alumina near bauxite mines, which will be predominantly in the LDCs, or that they will be required to do so by host countries. Expansion of primary aluminum capacity, however, is only economical in countries that have large potential sources of relatively low-cost energy, and requires the investment of very large amounts of capital.

Brazil is planning to use its hydroelectric potential to achieve self-sufficiency in aluminum and eventually to become an exporter of aluminum metal. Saudi Arabia is planning to use some of its abundant natural gas for aluminum smelting. Although smelters based on hydroelectric power in countries such as Guinea, Guyana, Brazil, and Costa Rica should have important cost advantages, the investment climate will play a major role in determining whether these and other countries will be able to find foreign partners and financing for aluminum smelting capacity.

Nickel

There is general agreement that world consumption of nickel will increase at a substantially lower rate than the historical 6.5 percent annual growth noted in Table A-1. Projected rates of growth in consumption given there vary rather widely, however, from 2.1 percent (Malenbaum, 1975–2000) and 3.5 percent (U.S. Bureau of Mines, 1975–2000) to 5.1 percent (World Bank, 1976–90).

Much of the planned increase in nickel producing capacity is located in developing countries, including Brazil, Colombia, Guatemala, Indonesia, and New Caledonia, and a substantial portion of this planned capacity expansion involves international mining companies. These companies have experienced political difficulties in a number of the countries where nickel resources are located, including the Dominican Republic, Guatemala and Indonesia. Yet, the diversity of supply sources together with the presence of substantial reserves in some OECD countries (including Canada, Australia and Greece) suggest that the development of adequate nickel producing capacity is not likely to become a serious problem during the remainder of the present century.[5] In fact, overcapacity may continue to be a problem for the next several years.[6]

The cost of nickel producing capacity per tonne of nickel is very high relative to that of other metals, and capital costs vary substantially with the location of the mine and the type of ore. For example, the cost for nickel laterite ore mines is 50 percent or more higher than for nickel sulphide mines. A recent U.N. study estimates the average capital cost of new nickel mining and processing capacity at $18,000 per annual tonne (in 1975 dollars).[7]

---

5 The United States has large nickel resources, but small production.

6 See, for example, "Nickel Product Price Peaked at $2.30 Per Pound Average for 1978," *American Metal Market* (April 3, 1978), p. 31.

7 *Minerals: Salient Issues*, Report of the Secretary-General, United Nations Economic and Social Council, E/C, 7/68 (New York, March 1977), Annex I, p. 6.

Assuming a 3 percent per annum growth in consumption, a rough estimate of additional required mine producing capacity for nickel in the period between 1977 and 2000 is about 625,000 tonnes at a capital cost of about $11.3 billion, or nearly $470 million per year. On the basis of present projected plans for capacity expansion, it would appear that at least 40 percent of the additional required capacity would be located in developing countries with an investment requirement of nearly $190 million per year.

Iron Ore

Because both the OECD countries and a number of developing countries, including Brazil and some of the African countries, are well endowed with iron ore reserves, the trend of exports of iron ore from the developing countries to the OECD countries is difficult to project to the year 2000. According to the Bureau of Mines, the United States is expected to reduce the proportion of its consumption supplied by imports from 36 percent in 1974 to 25 percent by 2000 as domestic production, principally of pellets from taconite, increases.[8] There are a number of announced projects for expanding Brazilian iron ore production, including pelletization, and that country's large reserves of high-quality iron ore have given it important competitive advantages. Brazil has a large state-owned mining company (CVRD), and there are several large iron ore joint ventures of international mining companies with Brazilian companies.

Projections in Table A-1 for the annual rate of increase in iron ore consumption range from 2.1 percent (Malenbaum, 1975–2000) to 2.8 percent (Bureau of Mines, 1973–2000) to 3.2 percent (World Bank, 1976–2000). Assuming a 3 percent per annum growth in world demand for iron ore between 1977 and 2000, additional required capacity may be estimated at nearly 600 million tonnes per year of iron content. At an estimated cost of $150 per annual tonne (in 1975 dollars), the capital requirements for the additional capacity would total nearly $90 billion. A rough estimate is that only 40 percent of these capital expenditures would be in the developing countries, or about $25 billion.

Tin

The OECD countries are heavily dependent on the developing countries—mainly Malaysia, Bolivia, Indonesia, and Thailand—for their supplies of tin. Tin production in the market economy countries reached a peak of 195,000 mt in 1972, but declined sharply thereafter. World consumption also declined sharply in 1974 and 1975, but rose by about 13 percent in 1976 over 1975, so that prices increased substantially. However, the demand for tin is expected to level off and to increase at an average rate of only 1.3 to 1.5 percent per annum during the remainder of the century, somewhat higher than the historical rate of growth (see Table A-1).

The tin mines are largely privately owned in Malaysia and Thailand while in Bolivia and Indonesia output is mainly under the control of state mining companies. Although Malaysia is currently the largest producer, with more than twice the output of any other country, Malaysian output is substantially below the 1972 production

---

8 Takeuchi, *et al.*, "Investment Requirements," p. 269; and an unpublished report prepared by the United Nations. Industry sources comment that this is too high an estimate except for remote projects requiring large-scale infrastructure. For most projects, new capacity might cost no more than $100 per annual mt. Thus, the above capital cost estimates for iron ore capacity may be overstated by as much as 25 to 40 percent.

peak, and, despite high prices during 1976–77, new investment in producing capacity has been low. The reasons given include high taxes and the discouragement of new investment and prospecting arising from the government's attitude toward foreign investment. The government is seeking to increase domestic participation in all foreign investments requiring all new foreign investment in the mining sector to have 70 percent Malaysian participation.[9]

Indonesia and Thailand each have more tin resources than Malaysia, and a Bureau of Mines report suggests that over the next two decades either country could become the world's largest tin producer.[10] Brazil and Zaire also have large tin resources and these countries could become very important sources of supply in the future. Changes in governmental policies and investment climate could affect the future global distribution of tin producing capacity.

The capital cost per metric ton of annual capacity of tin is estimated at $10,000 to $15,000 (in 1975 dollars).[11] Tin mine-producing capacity in the market economies in 1977 was estimated at about 192,000 tonnes metal content.[12] Assuming an increase in annual consumption of 1.5 percent to the year 2000, the capital outlays for the required additional mine capacity of about 80,000 tonnes through the year 2000 would be from $1.0 billion to $1.5 billion. The bulk of the additional capacity will probably be located in the developing countries.

---

9 John Thoburn, "Malaysia's Tin Supply Problems," *Resources Policy* (Vol. 4, No. 1), March 1978, pp. 31–34.

10 K.L. Harris, "Tin," in *Mineral Facts and Problems* (Bureau of Mines), p. 1140.

11 Takeuchi, *et al.*, "Investment Requirements," p. 269.

12 K.L. Harris, "Tin," in *Mineral Commodity Profiles* (U.S. Bureau of Mines, July 1978), p. 2.

APPENDIX D
Regional Cost Comparisons

Regional cost differences are determined in part by physical factors such as the size and grade of the ore bodies, their location in relation to existing infrastructure such as transportation facilities, power, water and housing, availability of trained and experienced workers, and their distance from markets. However, differences in physical conditions between countries may be overshadowed by differences in tax regimes, wages and employment conditions, government regulations, the cost of borrowing, and political risk. These latter factors may change substantially over a short period of time and often for the worse after an investment has been made. Thus, full economic costs of production in a developing country with large known reserves of high-grade ore may not necessarily be lower than costs in a developed country with a much less favorable geological environment.

A proper comparison of costs should be based on the per unit costs of production for new mines, including both operating and capital costs, with an allowance for return on equity sufficient to attract the investment. The opportunity cost of capital may differ considerably between a private international mining firm and a government enterprise that is not influenced by taxes and political risk.

The copper industry provides an example of cost differences among countries with different physical conditions. According to an article in *World Mining* (April 1976), costs at Sodimiza's Mushoshi mine were 90 cents per pound in 1975. The construction of the Tenke-Fungurume mine in Zaire had to be halted because of extremely high capital and prospective operating costs, despite the fact that the ore body was perhaps the highest-grade large copper ore body in the world, grading as high as 6 percent copper. In Zambia, where the ore grade is substantially higher than that of most U.S. and Canadian mines but where the mines are deeper and require more pumping of water, the total cost per pound of copper produced by Roan Consolidated Mines (RCM) was estimated by the *Engineering and Mining Journal* (January 1977) at approximately 60 cents per pound, and costs have risen substantially since then as a consequence of transportation and employment problems. RCM's costs were well above those of the lowest-cost U.S. mines in 1977.[1]

From the standpoint of the physical factors involved, Southern Peru Copper Company's Toquepala and Cuajone mines could be among the lowest-cost mines in the world, with operating costs in the range of 45 to 50 cents per pound for Cuajone in 1977.[2] Nevertheless, because of high taxes, including the Peruvian export tax, Charles Barber (Chairman of ASARCO which is the majority partner in SPCC) has estimated that a price of 97 cents per pound would be required to earn a 15 percent aftertax return on SPCC's investment in Cuajone in 1977. Furthermore, Barber stated that if Cuajone had been built at 1977 costs, an average copper price of $1.35

---

1 Several U.S. mines had costs under 60 cents per pound in 1977, and the U.S. national average was 62 to 65 cents per pound, of which 6 to 8 cents per pound was attributed to national pollution abatement requirements. See *Outlook for Development in the World Copper Industry*, CIPEC (Paris, November 1977), pp. 106–112. Much of CIPEC's data on U.S. costs was derived from an article by George W. Cleaver in *Metals Week*, June 24, 1977.

2 *Outlook for Development*, p. 40.

per pound would be required for SPCC to earn 15 percent return on its investment.[3]

1977 production costs for copper produced by Chile's government mining enterprise, CODELCO, were estimated at 48 cents per pound, and net operating costs were estimated at 34 cents per pound.[4] However, CODELCO's capital costs are quite low since the government took over the properties of foreign mining companies at book value during the 1966–72 period. Nevertheless, Chile should be regarded as a relatively low-cost producer of copper. The Bougainville mine in PNG and some of the Philippine mines have perhaps the lowest costs in the world, with costs per pound of copper in the range of 35 to 45 cents. In both cases, the low stated costs may be attributed to the relatively high gold content of the ore, the sales value of which is subtracted from the joint production costs to arrive at the calculated per pound "cost" of copper.[5] In the case of higher-cost mines, by-products or co-products such as nickel, molybdenum, gold, and silver play an important role. U.S. costs, which tend to average somewhat higher than Canadian costs, ranged from 55 cents per pound to over $1.00 per pound in 1977.[6]

In Canada, there is a wide range of costs per pound of copper from 40 to 70 cents per pound. The cost per pound of new mining operations in the United States in 1977 has been estimated at 91 cents per pound and rising to $1.04 per pound (in 1977 dollars) by 1980.[7]

---

3 Charles F. Barber, "Economics of New Supplies of Copper," *Mining Congress Journal* (March 1978), pp. 33–37. Barber's figures do not include the Peruvian government workers participation tax of 10 percent of earnings before taxes and depletion. If this tax is included, the price of copper would have to be $1.43 per pound in 1977 dollars in order for SPCC to earn 15 percent on its investment, assuming the mine had been constructed at 1977 prices at a cost of $1.2 billion.

4 "CODELCO: The Secrets of Success," *Copper Studies* (July 7, 1978), p. 1.

5 *Outlook for Development*, pp. 61–76.

6 *Ibid.*, p. 114.

7 *Ibid.*, p. 115.

Members of the British-North American Committee

Chairmen

SIR RICHARD DOBSON
President, B.A.T. Industries Limited, London

IAN MacGREGOR
General Partner, Lazard Frères & Co., New York, Honorary Chairman, AMAX Inc., Greenwich, Connecticut

Vice Chairmen

SIR ALASTAIR DOWN
Chairman, Burmah Oil Company, Swindon

GEORGE P. SHULTZ
President, Bechtel Corporation, San Francisco, California

Chairman, Executive Committee

WILLIAM I. M. TURNER, JR.
President and Chief Executive Officer, Consolidated-Bathurst Inc., Montreal, Quebec

Members

WILLIAM S. ANDERSON
Chairman of the Board, NCR Corporation, Dayton, Ohio

J.A. ARMSTRONG
Chairman and Chief Executive Officer, Imperial Oil Limited, Toronto, Ontario

JOSEPH E. BAIRD
President and Chief Operating Officer, Occidental Petroleum Corporation, Los Angeles, California

A.E. BALLOCH
Executive Vice President, Bowater Incorporated, Old Greenwich, Connecticut

ROBERT A. BANDEEN
President and Chief Executive Officer, Canadian National, Montreal, Quebec

*W.G. BARRETT
General Manager, Midland Bank Limited, International Division, London

SIR DONALD BARRON
Group Chairman, Rowntree Mackintosh Limited, York

DAVID BASNETT
General Secretary, General & Municipal Workers' Union, Esher, Surrey

CARL E. BEIGIE
President and Chief Executive Officer, C.D. Howe Research Institute, Montreal, Quebec

ROBERT BELGRAVE
Director, BP Trading Company Limited, London

JOHN F. BOOKOUT
President and Chief Executive Officer, Shell Oil Company, Houston, Texas

T.F. BRADSHAW
President, Atlantic Richfield Company, Los Angeles, California

JOHN F. BURLINGAME
Vice President and Group Executive, International and Canadian Group, General Electric Company, Fairfield, Connecticut

SIR GEORGE BURTON
Chairman, Fisons Ltd., London

SIR CHARLES CARTER
Chairman of Research and Management Committee, Policy Studies Institute, London

J. EDWIN CARTER
Chairman and Chief Executive Officer, INCO, Ltd., Toronto, Ontario

SILAS S. CATHCART
Chairman & Chief Executive Officer, Illinois Tool Works, Inc., Chicago, Illinois

SIR FREDERICK CATHERWOOD
Member of European Parliament and Chairman, Mallinson-Denny Ltd., London

HAROLD van B. CLEVELAND
Vice President, Citibank, New York, New York

JAN COLLINS
Chairman, William Collins & Sons, Glasgow, Scotland

*Became a member of the Committee after the statement was circulated for signature.

DONALD M. COX
Director and Senior Vice President, Exxon Corporation, New York, New York

RALPH J. CRAWFORD, JR.
Vice Chairman of the Board, Wells Fargo Bank, San Francisco, California

FRANK J. CUMMISKEY
IBM Vice President and President, General Business Group/International, IBM Corporation, White Plains, New York

JAMES W. DAVANT
Chairman of the Board and Chief Executive Officer, Paine, Webber, Jackson & Curtis Inc., New York, New York

DIRK DE BRUYNE
Managing Director, Royal Dutch/Shell Group of Companies, London

WILLIAM DODGE
Ottawa, Ontario

GEOFFREY DRAIN
General Secretary, National Association of Local Government Officers, London

JOHN DU CANE
Chairman and Managing Director, Selection Trust Ltd., London

DONALD V. EARNSHAW
Senior Staff Executive, Continental Can Company, Stamford, Connecticut

GERRY EASTWOOD
General Secretary, Association of Patternmakers and Allied Craftsmen, London

HARRY E. EKBLOM
Chairman and Chief Executive Officer, European American Bancorp, New York, New York

MOSS EVANS
General Secretary, Transport & General Workers' Union, London

J.K. FINLAYSON
Vice Chairman, The Royal Bank of Canada, Montreal, Quebec

GLENN FLATEN
First Vice President, Canadian Federation of Agriculture, Regina, Saskatchewan

ROBERT M. FOWLER
Chairman, Executive Committee, C.D. Howe Research Institute, Montreal, Quebec

GWAIN H. GILLESPIE
Senior Vice President—Finance, Heublein Inc., Farmington, Connecticut

MALCOLM GLENN
Executive Vice President, Reed Holdings Inc., London

GEORGE GOYDER
Hon. Secretary, British-North American Research Association, London

HON. HENRY HANKEY
Director, Lloyds Bank International Ltd., London

AUGUSTIN S. HART, JR.
Vice Chairman, Quaker Oats Company, Chicago, Illinois

G.R. HEFFERNAN
President, Co-Steel International Ltd., Whitby, Ontario

HENRY J. HEINZ II
Chairman of the Board, H.J. Heinz Company, Pittsburgh, Pennsylvania

ROBERT HENDERSON
Chairman, Kleinwort Benson Ltd., London

ROBERT P. HENDERSON
President and Chief Executive Officer, Itek Corporation, Lexington, Massachusetts

HENDRIK S. HOUTHAKKER
Professor of Economics, Harvard University, Cambridge, Massachusetts

TOM JACKSON
General Secretary, Union of Post Office Workers, Clapham, London

DONALD P. JACOBS
Dean, Graduate School of Management, Northwestern University, Evanston, Illinois

JOHN V. JAMES
Chairman of the Board, President, and Chief Executive Officer, Dresser Industries, Inc., Dallas, Texas

GEORGE S. JOHNSTON
President, Scudder, Stevens & Clark, New York, New York

JOSEPH D. KEENAN
President, Union Label and Service Trades Department, AFL-CIO, Washington, D.C.

TOM KILLEFER
Chairman of the Board and Chief Executive Officer, United States Trust Company of New York, New York

CURTIS M. KLAERNER
Vice Chairman, Commonwealth Oil Refining Company, San Antonio, Texas

H.U.A. LAMBERT
Chairman, Barclays Bank International Ltd., London

HERBERT H. LANK
Hon. Director, Du Pont of Canada Ltd., Montreal, Quebec

WILLIAM A. LIFFERS
Vice Chairman, American Cyanamid Company, Wayne, New Jersey

JAY LOVESTONE
International Affairs Consultant, AFL-CIO, Washington, D.C.

RAY W. MACDONALD
Honorary Chairman, Burroughs Corporation, Grosse Pointe, Michigan

CARGILL MacMILLAN, JR.
Senior Vice President, Cargill Inc., Minneapolis, Minnesota

J.P. MANN
Deputy Chairman, United Biscuits Holdings Ltd., Middlesex

A.B. MARSHALL
Chairman, Bestobell Engineering Products Ltd., Slough, Berks.

DENNIS McDERMOTT
President, Canadian Labour Congress, Ottawa, Ontario

WILLIAM J. McDONOUGH
Executive Vice President, International Banking Department, The First National Bank of Chicago, Chicago, Illinois

WILLIAM C.Y. McGREGOR
International Vice President, Brotherhood of Railway, Airline & Steamship Clerks, Montreal, Quebec

DONALD E. MEADS
Chairman and President, Carver Associates, Plymouth Meeting, Pennsylvania

PATRICK MEANEY
Group Managing Director, Thomas Tilling Limited, London

C.J. MEDBERRY III
Chairman of the Board, BankAmerica Corporation & Bank of America NT&SA, Los Angeles, California

SIR PETER MENZIES
Welwyn, Herts.

JOHN MILLER
Vice Chairman, National Planning Association, Washington, D.C.

DEREK F. MITCHELL
Chairman & Chief Executive Officer, BP Canada Limited, Montreal, Quebec

JOSEPH P. MONGE
Rancho Santa Fe, California

DONALD R. MONTGOMERY
Secretary-Treasurer, Canadian Labour Congress, Ottawa, Ontario

MALCOLM MOOS
Hackensack, Minnesota

KENNETH D. NADEN
President, National Council of Farmer Cooperatives, Washington, D.C.

WILLIAM L. NAUMANN
Former Chairman of the Board, Caterpillar Tractor Company, Peoria, Illinois

WILLIAM S. OGDEN
Executive Vice President, The Chase Manhattan Bank, N.A., New York, New York

PAUL PARÉ
President and Chief Executive Officer, Imasco Ltd., Montreal, Quebec

BROUGHTON PIPKIN
Chairman, BICC Limited, London

SIR RICHARD POWELL
Director, Hill Samuel Group Ltd., London

ALFRED POWIS
Chairman & President, Noranda Mines Limited, Toronto, Ontario

J.G. PRENTICE
Chairman of the Board, Canadian Forest Products, Ltd., Vancouver, British Columbia

LOUIS PUTZE
Director and Consultant, Rockwell International, Pittsburgh, Pennsylvania

BEN ROBERTS
Professor of Industrial Relations, London School of Economics, London

HAROLD B. ROSE
Group Economic Adviser, Barclays Bank Limited, London

DAVID SAINSBURY
Director of Finance, J. Sainsbury Ltd., London

WILLIAM SALOMON
Managing Partner, Salomon Brothers, New York, New York

A.C.I. SAMUEL
Director General, International Group of the National Association of Pesticide Manufacturers, London

NATHANIEL SAMUELS
Vice Chairman, Kuhn Loeb Lehman Brothers International, Chairman, Louis Dreyfus Holding Company, Inc. New York, New York

SIR FRANCIS SANDILANDS
Chairman, Commercial Union Assurance Company, Limited, London

HON. MAURICE SAUVÉ
Executive Vice President, Administrative and Public Affairs, Consolidated-Bathurst Inc., Montreal, Quebec

PETER F. SCOTT
President, Provincial Insurance Company, Ltd., Kendal, Westmoreland

ROBERT C. SEAMANS, JR.
Massachusetts Institute of Technology, Cambridge, Massachusetts

LORD SEEBOHM
Chairman, Finance for Industry, London

THE EARL OF SELKIRK
President, Royal Central Asian Society, London

JACOB SHEINKMAN
General Secretary-Treasurer, Amalgamated Clothing and Textile Workers' Union, New York, New York

LORD SHERFIELD
Chairman, Raytheon Europe International Company, London

R. MICHAEL SHIELDS
Managing Director, Associated Newspapers Group Ltd., London

GEORGE L. SHINN
Chairman, The First Boston Corporation, New York, New York

WILLIAM E. SIMON
New York, New York

*GORDON R. SIMPSON
Chairman, General Accident Fire and Life Assurance Corporation Ltd., Perth, Scotland

ARTHUR J.R. SMITH
President, National Planning Association, Washington, D.C.

LAUREN K. SOTH
West Des Moines, Iowa

---

* Became a member of the Committee after the statement was circulated for signature.

E. NORMAN STAUB
Chairman and Chief Executive Officer, The Northern Trust Company, Chicago, Illinois

RALPH I. STRAUS
New York, New York

JAMES A. SUMMER
Excelsior, Minnesota

HAROLD SWEATT
Honorary Chairman of the Board, Honeywell, Inc., Minneapolis, Minnesota

SIR ROBERT TAYLOR
Deputy Chairman, Standard Chartered Bank Ltd., London

A.A. THORNBROUGH
Deputy Chairman and Chief Executive Officer, Massey-Ferguson Limited, Toronto, Ontario

*J.C. TURNER,
General President, International Union of Operating Engineers, AFL-CIO, Washington, D.C.

SIR MARK TURNER
Chairman, Rio Tinto-Zinc Corporation Ltd., London

JOHN W. TUTHILL
President, The Salzburg Seminar, Cambridge, Massachusetts

W.O. TWAITS
Toronto, Ontario

MARTHA R. WALLACE
Executive Director and Director, The Henry Luce Foundation, Inc., New York, New York

RICHARD C. WARREN
Consultant, IBM Corporation, Armonk, New York

GLENN E. WATTS
President, Communications Workers of America, AFL-CIO, Washington, D.C.

W.L. WEARLY
Chairman, Ingersoll-Rand Company, Woodcliff Lake, New Jersey

VISCOUNT WEIR
Chairman and Chief Executive, The Weir Group Limited, Glasgow, Scotland

HUNTER P. WHARTON
General President Emeritus, International Union of Operating Engineers, Washington, D.C.

WILLIAM W. WINPISINGER
President, International Association of Machinists and Aerospace Workers, Washington, D.C.

SIR ERNEST WOODROOFE
Former Chairman, Unilever Ltd., Guildford, Surrey

Inactive Status
C. FRED BERGSTEN

*Became a member of the Committee after the statement was circulated for signature.

Sponsoring Organizations

The British-North American Research Association was inaugurated in December 1969. Its primary purpose is to sponsor research on British-North American economic relations in association with the British-North American Committee. Publications of the British-North American Research Association as well as publications of the British-North American Committee are available at the Association's office, 1 Gough Square, London EC4A 3DE (Tel. 01-353-6371). The Association is recognized as a charity and is governed by a Council under the chairmanship of Sir Richard Dobson.

The National Planning Association is a private, nonprofit, nonpolitical organization, founded in 1934, that carries on research and policy formulation in the public interest. NPA was founded during the great depression of the 1930s, when conflicts among the major economic groups—business, farmers, labor—threatened to paralyze national decision making on the critical issues confronting American society. It was dedicated, in the words of its statement of purpose, to "getting [these] diverse groups to work together . . . to narrow areas of controversy and broaden areas of agreement . . . [and] to provide on specific problems concrete programs for action planned in the best traditions of a functioning democracy." NPA is committed to the view that the survival of a functioning American democracy under the increasingly rigorous conditions of the 20th century requires not only more effective government policies but also preservation of private economic initiative and the continuous development by the major private groups themselves of a consensus on how to cope with the problems confronting the nation at home and abroad.

NPA works through policy committees of influential and knowledgeable leaders from business, labor, agriculture, and the professions that make recommendations for dealing with domestic and international developments affecting the well-being of the United States. The research and writing for these committees are provided by NPA's professional staff and, as required, by outside experts. In addition, NPA's professional staff undertakes a wide variety of technical research activities designed to provide data and ideas for policy makers and planners in government and the private sector. These activities include the preparation on a regular basis of economic and demographic projections for the national economy, regions, states, and metropolitan areas; policy-oriented research on national goals and priorities, employment, manpower needs and skills, health, energy, environment, science and technology, and other economic and social problems confronting American society; and analyses and forecasts of changing international realities and their implications for U.S. policies.

NPA publications, including those of the British-North American Committee, can be obtained from the Association's office, 1606 New Hampshire Avenue, N.W., Washington, D.C. 20009 (Tel. 202-265-7685).

The C.D. Howe Research Institute is a private, nonpolitical, nonprofit organization founded in January 1973, by the merger of the C.D. Howe Memorial Foundation and the Private Planning Association of Canada (PPAC), to undertake research into Canadian economic policy issues, especially in the areas of international policy and major government programs.

HRI continues the activities of the PPAC. These include the work of three established committees, composed of agricultural, business, educational, labor, and professional leaders. The committees are the Canadian Economic Policy Committee, which has been concentrating on Canadian economic issues, especially in the area of trade, since 1961; the Canadian-American Committee, which has dealt with relations between Canada and the United States since 1957 and is jointly sponsored by HRI and the National Planning Association in Washington; and the British-North American Committee. Each of the committees meets twice a year to consider important current issues and to sponsor and review studies that contribute to better public understanding of such issues.

In addition to taking over the publications of the three PPAC committees, HRI releases the work of its staff, and occasionally of outside authors, in four other publications: *Observations,* six or seven of which are published each year; *Policy Review and Outlook,* published annually; *Special Studies,* to provide detailed analysis of major policy issues for publication on an occasional basis; and *Commentaries,* to give wide circulation to the views of experts on issues of current Canadian interest.

HRI publications, including those of the British-North American Committee, are available from the Institute's offices, 2064 Sun Life Building, Montreal, Quebec H3B 2X7 (Tel. 514-879-1254).

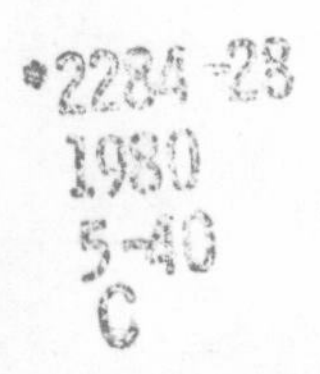